Generic Drugs Formulation Manual

Basic Principles of New Products Development

2nd Edition

by Francisco Delatorre Quiñónez.

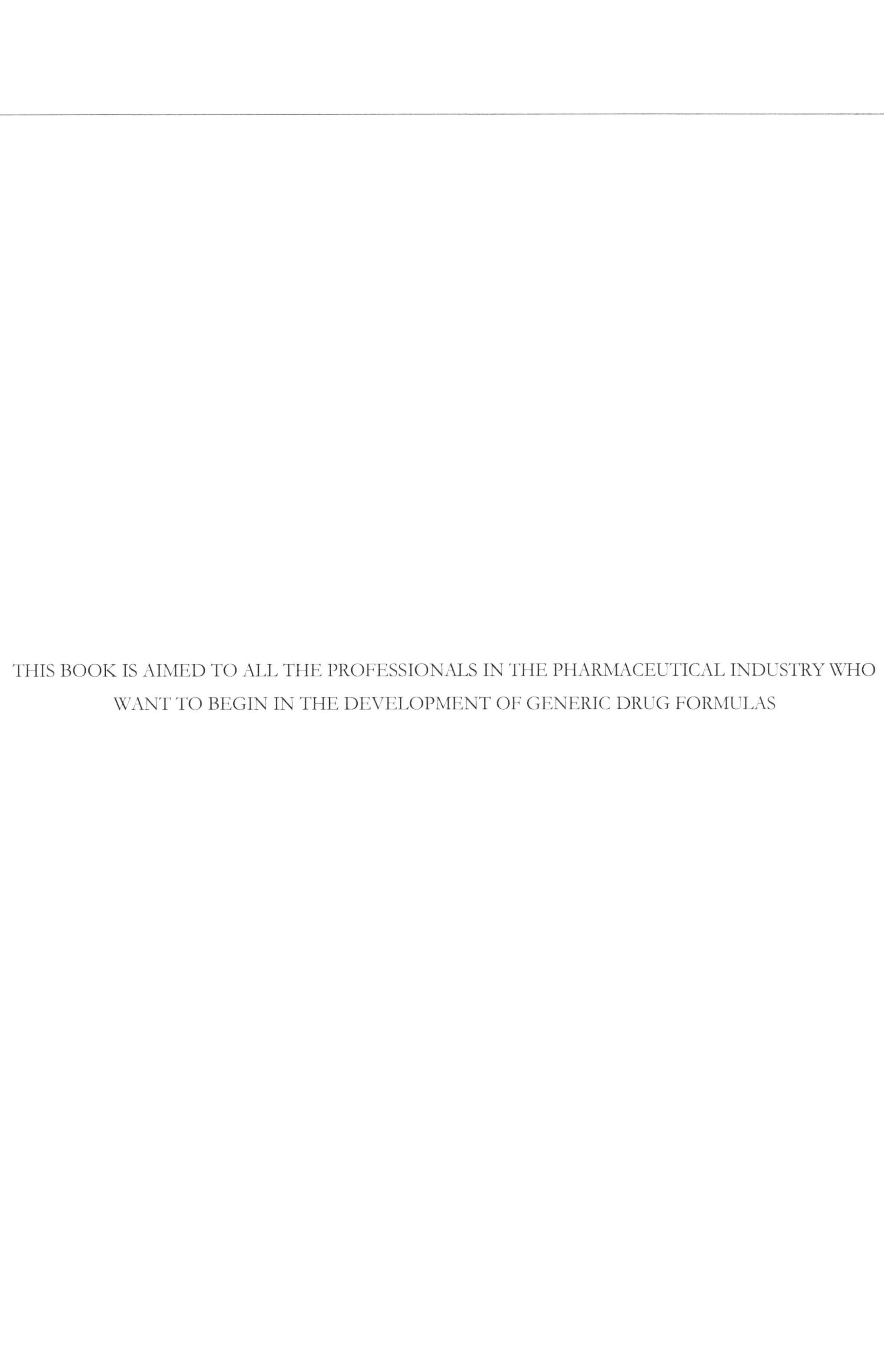

THIS BOOK IS AIMED TO ALL THE PROFESSIONALS IN THE PHARMACEUTICAL INDUSTRY WHO WANT TO BEGIN IN THE DEVELOPMENT OF GENERIC DRUG FORMULAS

FRANCISCO DE LA TORRE QUIÑÓNEZ, novel writer on Pharmaceutical Industrial technology comes with his second book on drug development: **"Generic Drugs Formulation Manual: Basic Principles of New Product Development"**; bringing us simple formulas of considerable commercial interest which are a step beyond in comparison to other works of this kind.

THIS PAGE INTENTIONALLY LEFT BLANK

ACKNOWLEDGMENTS

First of all, I thank God and his son: Jesus Christ for every step they allow me to take in my life.

I thank my children: Edward De La Torre Paredes and Rubí De La Torre Paredes, for giving my existence a light of joy every day.

To my mother Raquel Mercedes Quiñónez Jínez for her dedication and effort on a day-to-day basis.

To Sonia de los Ángeles Quiñónez Jínez (+), my second mother, who, being a Chemical Engineer, introduced me to Sciences and Industry.

To my grandfather: Juan Vitaliano Quiñónez Albán (+) who every Thursday visited our home bringing us gifts as a result of his daily work. I also thank his sister Águeda Quiñonez Albán (aunt Angelita), who was the person who raised my mother and my aunt Sonia.

To Mercedes María Jínez Sánchez (+), my grandmother, who gave us her delicacies and Ecuadorian gastronomic skills; and who also saved my life when I was 6 years old.

To my grandparents Andrés De La Torre Gallegos (+) and María Ramírez (+) who were always closed to me despite of not living in my house.

To my former boss Ms. Karen Montjoy, MSc. who gave me her support at the beginning of my career.

To Dr. Erika Chavez (Erika con K) who always gave me her support in my job and also let me work opened to creativeness and innovation.

To Dr. Walter Mariscal who advised me to **"cross the Rubicon"** and thanks to him I was able to write everything I learned in my professional life, but also this encouraged me to become a writer in other disciplines, such as : General Fiction, Theology, Christian Eschatology, Science Fiction, Novels (in the future) (different literary genres). Books that I have written with my pseudonym: Juan Vitaliano Quiñónez Albán (in honor of my maternal grandfather).

To Dr. Soraya García-Mariscal who taught me to develop theoretical frameworks and also gave me her support both in my student life and in my career.

To the Pharmacist-Chemist community of Guayaquil (Ecuador), and also to the International community, who provided me their support in my Ebook (in Spanish): **MANUAL DE FORMULACIÓN DE SOLIDOS ORALES** published in 2022 through Amazon's KINDLE platform, who also supported my book: **Manual de Formulación de Medicamentos: Introducción al Desarrollo de Fórmulas (Spanish Edition)** published on 2023.

To all who believed and still believe in me. And to those who didn't, I also thank.

And in a very special way I thank you, Dear Reader, who has acquired this work, which I hope you enjoy.

PART I

GENERAL CONCEPTS OF SOLID, LIQUID, SEMI-SOLID AND SEMI-LIQUID PHARMACEUTICAL DOSAGE FORMS

1. GENERAL CONCEPTS OF SOLID ORAL PHARMACEUTICAL DOSAGE FORMS FOCUSED ON NEW PRODUCTS DEVELOPMENT

Tablets are a solid dosage form consisting of one or more active ingredients grouped by means of suitable excipients. They are normally prepared by compression or another previously known result process.

Tablets are mixtures of active ingredients and excipients, powders or granules pre-compacted in a single solid dosage form for mostly oral administration.

Within the excipients we find: diluents, binders, granulators, lubricants; additionally we find disintegrating agents that help break down the tablet in the gastrointestinal tract; as well as sweeteners or flavorings agents, if the case merits it; with the purpose of improving the organoleptic properties. A tablet formulation can include a polymer coating either to mask taste or to protect the active ingredient from degradation; to control the rate of release of the active ingredient, or to improve the tablet stability during storage and transportation.

Tablets are formulated for site-specific delivery; they are usually administered orally, but can also be sublingual or vaginal.

Types of Tablets. –
In general we can classify tablets as: pills, caplets and orodispersible tablets (known as ODT or Orally Dissolving Tablets).

Pills. – Initially presented as solid, small, round pharmaceutical forms, intended for oral administration. Evidence of its use has been found for more than 2,000 years ago. The first pills found in an Italian shipwreck near Pozzino, were made of zinc carbonate (hydrozincite and smithsonite); and were used for the treatment of eye irritation, dating their use from about 140 B.C.E.

Caplet. - Solid oval shaped similar to a capsule. It is considered the most efficient method of medication, and this is the reason why manufacturers of drugs such as OTC pain relievers (Over the Counter or free sale) emphasize the strength of this pharmaceutical form.

Orodispersible tablets (called as ODT or Orally Dissolving Tablets). – They can dissolve in the oral cavity without having to be swallowed; or, failing that, they can be placed in water (the medium in which they are dissolved), so that once dissolved they can be ingested (not to be confused with effervescent tablets).

1.1. TABLETS

Tablets Properties

The tablets can be manufactured in any shape, although the requirements of the patients and the most widely distributed tableting machines give preference to the manufacture of round, oval or capsular tablets. Tablets in other shapes have been made, however there is some reluctance to make new tablet designs.

The diameter and shape of the tablet are determined by the tableting machine. A tablet press is a mechanical device used to compress powders, granules, or granulated powders into tablets of uniform weight and size. To form a tablet, the granule must be inserted into a cavity consisting of two punches and a die. The punches are pressed with such force that they combine the powders (granules or granulated powders) into a uniform, compact dosage form.

In tablet compression the first step of a typical operation involves lowering the lower punch into the die creating a cavity in which the granule will go. The exact depth of the bottom punch can be controlled to measure the amount of powder that will fill the cavity. The excess is removed from the top of the matrix. Then the upper punch falls firmly in contact with the powders. The compression force is provided by high pressure compression rollers, which fuse the granulated material into a hard tablet. After compression the lower punch is raised to release the tablet. Being a correct setting of the tableting machine a critical point in order to guarantee the properties of this pharmaceutical form.

In the tablet press machines, there are basically two types: the single-stroke tablet press and the rotary tablet press (multi-station). Higher speed tablet presses come with a rotating turret or platform that holds a set number of punches. As these rotate around the turret, the punches have contact with cameras that control the vertical position of the punch.

Punches and dies are usually designed for each application; they can be manufactured in a wide variety of shapes, sizes or can be customized according to customer requirements. Depending on tablet size, shape, material, and configuration, a current tablet press (as of 2023) can produce around 100,000 tablets per hour.

The diameter and shape of the tablet are determined by the matrix and two punches (one upper and one lower) of the tablet press machine; this is known as a **station**. During manufacturing, the height of the tablet is determined by the amount of granules, the position of the punches, and their relationship during compression.

Once the height of the tablet is determined, the corresponding pressure applied during compression can be measured. The resulting tablets must be hard enough to withstand shipping and handling until delivery to the consumer (or patient), and likewise must be friable enough to disintegrate in some portion of the gastrointestinal tract.

The main measure of the mechanical strength of a tablet is the determination of its hardness. The equipment used for this determination are the hardness testers. Tablet hardness analysis is a technique used by the pharmaceutical industry to measure the breaking point and structural integrity of the tablet. The breaking point of the tablet is based on its shape; it is similar to a friability test, but it is not the same.

For the determination of the hardness, the technician (or duly trained operator) aligns the tablet in a repeatable way, and it is crushed between two structures. The first structure (the one in motion) applies a continuous force until the tablet begins to break. It is at this point that the hardness is read by the equipment.

The S.I (International System of Units) unit of force that is used to measure tablet hardness is the Newton; another unit is the kilopond (KP or kilogram force), although its use is practically discontinued, except in certain equipment in pharmaceutical companies that usually have been on the market for several decades.

In the manufacture of tablets another parameter of interest is the friability. Friability describes the tendency of a solid to break into parts under impact or contact. Tablet friability determination is a technique used to measure the durability of tablets during transit. This test involves the repeated launching of a sample of tablets for a specified time using a rotary fixture with a deflector to change its direction. The result is measured based on the broken tablets and the calculation is made based on the percentage of mass lost at the end versus the initial weight of the tablets.

Tablet manufacturing overview

In the tableting process the main thing is to ensure that the appropriate amount of active ingredient is distributed on each tablet. For this reason, all the ingredients must be well mixed. If somehow a sufficiently homogeneous quantity of the components cannot be obtained by a simple mixing process, the ingredients must be previously compressed, in order to ensure a good distribution of the active component in the final tablet.

Two basic methodologies are used to granulate the powders to be used in the tablets:
- Wet granulation and,
- Dry granulation process.

Those powders whose granule is of the correct size and that do not require prior granulation (whether by wet or dry granulation process) can be compressed by direct compression.

<u>Wet Granulation</u>

Wet granulation is a process that uses a liquid and a binder to agglomerate the mixture. The amount of liquid must be controlled appropriately, so that two things are avoided: a) over-wetting which can make the granules too hard, and, b) under-wetting can make the granules too soft and friable. Aqueous solutions have the advantage of being safer to handle than solvent-based systems, but these aqueous solutions are not suitable for drugs degraded by hydrolysis.

<u>Wet Granulation General Procedure</u>

1. The active ingredient and the excipients are weighed or measured depending on each case.
2. These substances are then placed in suitable mixing equipment.
3. The "wet granule" is prepared by adding a binder solution (depending on the case) to the product that is in the mixing equipment. After adding the binder solution, mixing continues in the equipment. It may be, and depending on the process that the binder goes into the mix inside the equipment and for the formation of the "wet granule" the solvent is added (whether this is water or alcohol).
4. Once the components are mixed, the bulk is placed in a powder granulator that has a mesh of adequate size according to the process.
5. After passing through the powder granulator, the bulk is placed in metal trays covered on their top with white wax paper (aka white baking paper) (this is for oven drying).
6. Once all the contents of the mixer have been discharged (the same that has already passed through the mesh of the powder granulator), the metal trays are placed in the oven and the temperature is set.
7. Once dry, the product hardens considerably, for this reason it must once again pass through the powder granulator (with a suitable mesh depending on the process); this, until obtaining a granule of uniform size, which can be mixed in the equipment together with other excipients prior to compression.

<u>Dry Granulation Process</u>

This procedure is based on a previous mixture to which lubricant has been added, in a small amount to what is stipulated in the total formula, before the bulk is placed in the tablet press.

The first thing that is sought in this way is to obtain some tablet slugs (large tablet) (dry granulation is also called as slugging) (the tablet slugs are tablets that may (or may not have) the same shape as the final product, as well as they may have the same, similar or less weight than the tablet to be finally obtained).

The successive breaking, granulation and tableting of the slugs (tablet slug) <u>seek to obtain a pre-compressed granule</u> as less dusty as possible, in order to obtain a resulting product with the appropriate characteristics.

Dry granulation is given by a continuous and successive compression and granulation process (compress, granulate; break the granules; compress, granulate, break the granules again).

It is important that after any mixing within this process of granulation - tableting - breaking of the tablet slug; proceed with a minimum, but adequate lubrication of the bulk. This in order to ensure that the powder does not stick to the punches of the tablet press. This small amount of lubricant during the formation of the tablet slugs prevents damage to the punches, allowing an adequate work flow of the tablet press. A mixture without lubricant will not tableting, but also the lubricant absence will cause wear, damage of the punches and their dies; and not only this, but forcing the machine to suffer considerable damage due to a possible over heat of engines. The key point to consider in this step is that the lubricant does not interfere with the dissolution of the active ingredient during laboratory analyses; this at the same time that during the successive compressions-breaks of the granule, the lubricant allows obtaining an adequate tablet slug and then a finished product with a suitable gloss or visually presentation, according to the specifications.

Tablet coating

After the tableting process many tablets are coated. In the past sugar coated tablets were popular (those similar to dragees), currently the use of polymers and polysaccharides with pigments (usually included) are the preferred materials for coating.

A coating on a solid dosage form must be stable and strong enough to maintain product specifications during storage and shipping. An adequate coating guarantees: a) that the tablets do not stick to each other, b) the absence of irregular edges or any layer that hides the logo or the tablet score (when applicable).

The coating of a tablet is necessary for those that have an unpleasant taste. Proper finishing of the coated tablet can make it easier to swallow. The coating is also useful to extend the stability of the tablet or to protect some of its components that could be sensitive to oxidation, humidity or some other type of degradation.

An enteric coating (the release and absorption of the active ingredient from the tablet or capsule occurs in the small intestine) is recommended for those active ingredients that are sensitive to acidity or irritates the stomach. The selection of a suitable coating agent must be carried out based on the well-known technical literature. There is no chance for improvisation or experimentation, especially if we are making a product whose focus will be commercial. A stomach acid resistant coating will dissolve and absorb in a lower acid intestinal area.

The enteric coating is also used for those oral solids that could be negatively affected in the time it takes to reach the small intestine; where it will be absorbed. Coatings are generally chosen to control the dissolution rate of the drug in the gastrointestinal tract; there are drugs whose absorption is better in certain sections of the digestive system. If the section to which we refer is the stomach, the coating to be selected must be one that allows dissolution in an acid medium. On the other hand, for those whose absorption rate is better in the colon, a slow-absorbing, acid-fast type coating would be the most suitable to ensure that the tablet reaches this section of the digestive tract, before dispersing.

In general, there are two types of coating machines used in the industry: coating drums and automatic coating equipment. The coating drums are used in those tables coating on the basis of sugar; instead, automatic equipment is used for all kinds of coatings, and can be equipped with a remote control panel, a dehumidifier and a suitable dust collector. For those coatings that contain isopropyl alcohol an explosion proof coating system design is recommended.

The motivations for tablet coating are diverse; from commercial or "aesthetic" (gloss, color) reasons, improvement of the product stability (protection from light, humidity, air); and to facilitate swallowing of the oral pharmaceutical form in those products (i.e antibiotic) which could be bitter (even emetic) upon first contact with the oral cavity during drug administration. The modification of the release behavior of the active ingredient can be caused due to a technical-chemical-commercial decision to coat the tablet. Generally the coated surface in a tablet could have a thickness of between about 20 to 100 microns.

Coating formulations can usually include the following elements among their components: a polymer, a plasticizer, coloring agents, an opacifier, a solvent, and a vehicle (depending on the case).

1.   Polymer

Within this group we have those derived from cellulose (e.g. cellulose ethers) and acrylic polymers or copolymers. Among the polymers we have high molecular weight polyethylene glycols, P.V.P (polyvinylpyrrolidone), waxes and polyvinyl alcohol. Frequently a polymer is dissolved in a suitable solvent, such as water, or in a non-aqueous solvent. Despite this, some water-insoluble polymers are available for use with aqueous systems; These materials have an interesting application in controlled release coating methods, the best known being true latexes and pseudo latexes.

1. a. Coating film formation mechanism

The mechanism of formation of a film for an aqueous polymer dispersion has been reviewed by several authors. When wet, the polymer appears as a number of discrete particles, which bind, deform and coalesce (or combine) thus giving rise to the formation of the film. In this scenario, the water will be lost as water vapor and the polymer particles increase their proximity to each other. Complete coalescence will occur when adjacent particles have the ability to diffuse into each other.

I.   Most Common Polymers Used in Coatings

A.   Cellulose

a.1. Hydroxyethylcellulose (HEC). Soluble in water, insoluble in organic solvents.

a.2 Hydroxypropyl cellulose (HPC). Substance soluble in both aqueous and alcoholic solvents. His films however tend to be sticky and weak; this slow down its coating property. It is used with other polymers in order to increase adhesion.

B.   Acrylic polymers

This group includes polymers with diverse functionalities.

b.1. Amino methacrylate copolymer. - Polymer basically insoluble in water, it can dissolve in solutions below pH 4. In neutral or alkaline environments, its films reach solubility by swallowing and the permeability of these films increases in aqueous media. Formulations designed for a traditional coating can be modified to increase its expansion and permeability by incorporating water-soluble materials such as cellulose ethers and starches; this in order to ensure a complete dissolution/disintegration of the coating film. This copolymer is supplied both as a powder and as a concentrated isopropyl alcohol/acetone solution, which can be diluted with solvents such as ethanol, methanol, acetone, and methylene chloride.

Materials such as: talc, magnesium stearate, or similar, can be added to the coating formula; this in order to reduce the sticky nature of the polymer.

Furthermore, the difference between a polymer used for a simple (non-functional) coating and one used for a modified coating does not have a classification or categorization within these divisions; several polymers can fulfill both functions.

II. Modified Release Polymers

A. (Meth) Acrylic Ester Copolymers

Relatively similar in structure to methacrylic acid copolymers, they are neutral and insoluble substances above physiological pH, but have the ability to swell and become permeable to water and dissolved substances, thus becoming useful in coating modified-release solid forms. Incorporated hydrophilic materials such as soluble ethers of cellulose, or PEG (polyethylene glycol) allow the formulation to be modified until the desired characteristic is found. Substances like Eudragit (either as Eudragit RS and Eudragit RL) can be mixed and matched to achieve a good release profile. The strong permeability of Eudragit RL, and its slight or slight retarding capacity, make its films suitable for rapidly disintegrating tablets. In aqueous coatings, one latex of each polymer is available, as is the case with Eudragit NE 30 D; polymer that is additionally used in non-functional immediate release coatings, for those formulations that have large amounts of water soluble materials.

B. Ethylcellulose (EC)

Cellulose ether resulting from a reaction of ethyl chloride with an appropriate alkaline cellulose solution. Beyond its extensive use in sustained-release coatings, EC is useful in organic solvent-based coatings in a blend with other cellulosic-type polymers, especially HPMC (hydroxypropylmethyl cellulose). Ethylcellulose gives an extra gloss to the tablet surface and in many ways is an ideal polymer for controlled release coatings; also presenting an absence of odor and taste and a high degree of stability to light and heat, not only under physiological conditions, but also under normal storage conditions. The solubility of EC to common solvents used for coating is good; however, this characteristic is not of great importance when faced with water-dispersible presentations, which have been specially designed for modified release. This polymer has tended to be disused by itself, but has been combined with other secondary polymers such as hydroxypropylmethyl cellulose or polyethylene glycol; substances that give EC a more hydrophilic nature to the film, altering its structure by virtue of pores and channels that allow the diluted drug to diffuse easily.

2. Enteric Polymers

These are the type of polymers whose design resists the natural acidity of the stomach, thus allowing the dissolution of the drug in the duodenum.

a. Cellulose acetate phthalate (CAP)

Insoluble in water, alcohol, and chlorinated hydroxycarbons. A pseudolatex version is available as a dry powder (Aquateric) for reconstitution in water for use in aqueous procedures. Due to their chemical constitution, enteric coatings of this type tend to present a variable degree of stability.

b. Polyvinyl Acetate Phthalate (PVAP)

This substance has the following solubility characteristics:

- 50% soluble in methanol

- 30% soluble in Methanol/Methylene Chloride

- 25% soluble in 95% Ethanol

- 30% soluble in Ethanol/water 85:15

A water dispersible form (Sureteric) is available for aqueous coatings.

c. Shellac

Purified resin from the secretion of the insect Laccifer Lacca, native to India and other parts of the Far East. Shellacs can be modified to fit specialized needs. Shellac is insoluble in water, but may show solubility in alkaline solutions or be sparingly soluble in hot ethanol. Shellac modifications often cause distribution problems for this raw material and variation in its quality.

d. Methacrylic Acid Copolymers

Due to the fact that they have free carboxylic groups, they are used in coatings of enteric materials, forming salts with alkalis and having an appreciable solubility at a pH above 5.5. Of the two organic solvent soluble polymers, Eudragit S100 has a lower degree of substitution with carboxyl groups and consequently dissolves at a higher pH than Eudragit L100. Used in combination, these materials are capable of providing films with a useful range of pH over which their solubility will occur. The addition of pigments and other substances to dispersible Eudragit forms, such as L30D and L100-55, should be done according to the manufacturer's recommendations, in order to prevent coagulation of the coating solution.

3. Plasticizers

Low molecular weight materials with the ability to alter the physical properties of a polymer, thus making it more useful in its function as a coating agent (softer and more flexible). The mechanism of action of a plasticizer is generally described in that its molecules are interposed between the polymer bonds and thus the polymer-polymer bonds are broken. This action is facilitated because the polymer-plasticizer bonds are considered stronger; and under this model it can be visualized how a plasticizer is capable of transforming a polymer into a more flexible material. Experimentally, the effect of a plasticizer on a polymer system can be demonstrated in many ways; a fundamental property of a polymer that can be determined by various techniques is the determination of the Glass Transition Temperature (Tg); this is the temperature at which a polymer changes from its hard, glassy form to a softer, more rubbery form. The plasticizer reduces the Glass Transition Temperature.

Plasticizers can be classified into three groups

1. Polyols
a. Glycerol (glycerin)
b. Propylene glycol
c. PEG polyethylene glycols (usually those from grade 200 to 6000)

2. Organic Esters
a. Phthalic acid esters (PAEs) (diethyl-; butyl-)
b. Dibutyl sebacate
c. Citric acid esters (tri ethyl-, triethyl acetyl-, tributyl acetyl-)
d. Triacetin

3. Oils/Glycerides
a. Castor Oil
b. Acetylated monoglycerides
c. Fractionated Coconut Oil

4. Opacifiers/colorants

Opacifiers and dyes allow product identification by both the manufacturer and patients. In the case of patients it can be useful for those who take multiple medications; it even helps manufacturers reduce a potential risk of counterfeiting.

Classification. -

Organic Dyers and lacquers

In this group we find raw materials such as Sunset Yellow, Patent Blue V, Quinoline Yellow, etc. Regarding tablet coating, its use is restricted due to its solubility in water. However, its water-insoluble complexes with hydrated alumina, known as lakes, are widely used for coatings.

b. Inorganic Dyes

Light stability is an important characteristic demonstrated by these materials, some of which have useful opacifying ability, e.g. titanium dioxide. Another great advantage of inorganic dyes is their wide regulatory acceptance, making them useful for multinational companies in order to standardize international formulas.

c. Natural dyes

These are a chemically and physically diverse group of materials. The description "natural" is slightly vague, because some of these products are chemically synthesized and not so much extracted from a natural source; the term generally applies to those materials as "nature identical", which in many ways would be more descriptive. Generally, these types of dyes are not light stable like other groups of dyes. However, they have a regulatory advantage which gives them high acceptability.

5. Solvents

These materials have a necessary function in that they provide the means of attachment between the coating materials to the tablet surface.

The most common used solvents are:

- Water

- Alcohols

- Ketones

A prerequisite for the solvent would be that it has to interact with the selected polymer; this is needed because a polymer-solvent interaction allows optimization of adhesion and mechanical strength in the coating.

2. GENERAL CONCEPTS OF ORAL LIQUID, LIQUID TOPICAL, ORAL SEMI-LIQUID AND TOPICAL SEMI-LIQUID PHARMACEUTICAL FORMS FOCUSED ON DRUG DEVELOPMENT

1. Liquid Dosage Forms

Pharmaceutical dosage form that contains a solid or a liquid (or several of these as the case may be) in one or several suitable solvents that are in a greater concentration or proportion than the active ingredient. They are mostly aqueous, alcoholic or hydroalcoholic.

Within these pharmaceutical forms we will find in general:

a. Oral liquid dosage forms

For oral administration, as their name indicates, they usually contain (depending on the properties of the active ingredient), some flavoring, sweetener and/or taste masker. They can be for oral ingestion or for some oral topical treatment.

They are homogeneous mixtures, rarely viscous, and may have some coloring for commercial purposes.

The oral liquid dosage forms generally lacks of solubilizing polymers, and if it had some type of polymer, as it happens with the preparations that have CARBOWAX 1450 (PEG 1450), then, this substance does not help with the suspension, dispersion or emulsification of the components of the formula, but, rather, the polymer allows the dissolution of an active ingredient with little or no solubility in aqueous solvents at room temperature, as occurs when preparing a syrup containing PEG 1450 and acetaminophen.

b. Topical liquid dosage forms

For external use, they may contain in their composition (in addition to the active ingredient and depending on the formulation) some emollient, exfoliant, healing agent or cooling agent (such as menthol), of performance already known in the classic galenic pharmacy. To differentiate it from a semi-liquid preparation we can mention that they do not contain gums or emulsifying polymers (at least not in a proportion like emulsions and suspensions), but rather are formed by simple mixtures (through a "solubility game" see which solid or liquid is soluble in another before final mixing). Additionally, topical liquid preparations (and it happens in the same way with oral liquids) DO NOT require the help of any homogenizing equipment (such as the Gaulin homogenizer) to achieve the combination of its components.

c. Oral semi-liquid dosage forms

They contain in their composition some emulsifying or suspending agent that allows a homogenization of the components that present little solubility.

We find within this category suspensions and emulsions. Suspensions are those pharmaceutical forms that have one or more solids, which would normally be insoluble in the formulation vehicles, dispersed with the help of some suspending agent, they are mostly viscous and must remain homogeneous when left to rest for a long time. Emulsions, on the other hand, consist of those immiscible or insoluble substances of a liquid or semi-liquid nature (in some preparations they might be semi-solid) found in combination with other excipients (or with only a vehicle or solvent) in which they would normally be immiscible. It has in its composition an emulsifying agent (which allows the combination of oily and non-oily substances), and at the time of its preparation it may or may not use a homogenizer (Gaulin type or another appropriate one) to maintain homogeneity in the product. It must additionally be stable when left at rest, avoiding to a minimum (or 99.99%) the separation of its phases (at rest).

d. Topical semi-liquid dosage forms

For external use, they contain in their composition emulsifying or suspending agents depending on the preparation. Likewise, and being topical, they may contain emollient, astringent or refreshing agents (depending on the formulation). Of fluid viscosity (differentiation from low-viscosity creams and ointments), they are an intermediate pharmaceutical form between liquids and semisolids. Its preparation, mostly hot, often implies the simultaneous use of emulsifying waxes or suspending gums in the same formula. For its preparation and final product, a homogenizer may be required (through which it is processed in its most fluid state).

2.1. SYRUPS

Syrups are oral preparations that seek a pleasant taste (an issue that could be complicated depending on the active ingredient). Many of its formulations contain sugar (sucrose), however, this must be carefully considered depending on the population target (group to which the product is directed); since there could be patients with a tendency to diabetes, for whom this type of formulation would be totally contraindicated, preparations sweetened with combinations of sucralose and/or sodium saccharin (or aspartame) should be designed instead.

The syrups are homogeneous, transparent (free of opacity and turbidity) and generally colored. Designed to be used in children and adolescents, they can be a consumption option for those active who require little dosage or in certain cases for those adults who find it difficult (due to various factors) to swallow whole tablets.

2.2. ORAL SOLUTIONS

Homogeneous, colored or transparent, with no turbidity (must have minimal or no visible opacity); the solutions have a much lower viscosity compared to a syrup. Oral solutions are characterized by their fluid presentation and are mostly used in oral preparations such as mouthwashes, related products; or in oral rehydration medications. They are generally packaged in transparent PET bottles (depending on the stability and characteristics of the product), and they tend to be pharmaceutical forms whose commercial presentation (in what regards to all of their content), is completely consumed during a not very long period, unlike a syrup which its content may be left over after the treatment indicated by the Physician is finished, or in some cases when the symptoms cease.

3. GENERAL CONCEPTS OF SEMI-SOLID PHARMACEUTICAL DOSAGE FORMS APPLIED ON PRODUCT DEVELOPMENT

3.1   TOPICAL DOSAGE FORMS (CREAMS, OINTMENTS, GELS)

3.1.1   CREAMS

Semisolid pharmaceutical form, generally topical, containing in its composition oily and non-oily substances emulsified by some appropriate agent, and may also have dissolved or suspended solids. They are often fluid, but with a viscosity (depending on the product) somehow higher than semi-liquid pharmaceutical forms. Usually packaged in aluminum tubes, bottles with a dispenser and rarely in pots or knobs, they can have various applications ranging from therapeutic use to cosmetic use.

3.1.2   OINTMENTS

Topical semi-solid pharmaceutical dosage form generally appears oily, without any flow property, or poorly fluid (compared to a cream). Most of the times used in pharmaceutical industry, they usually come in tubes or jars. Its preparations require a heating process, and they are rarely manufactured at room temperature. There are manufacturing processes in which is required a period of 'conditioning' (some sort of a wait time) before packaged in its primary packing; this implies that a manufacturing of these sorts of products must be done with a considerable anticipation before they are marketed.

3.1.3   GELS

Pharmaceutical dosage form that contains a gelling agent (it could be carbopol or carbomer 940) in such a proportion as to have its final or commercial appearance. For this type of preparation, the prior hydration time of the gelling agent (gum or polymer) is important, as well as adequate neutralization of the polymer (which is usually acidic most of the time) against an alkalizing agent (such as triethanolamine).

Most of them are topical (except for multivitamin complexes, which we will not discuss in this text), they may contain substances that would normally be insoluble in their vehicles, and may require the use of solubilizing agents or emulsifying agents for essential oils or some other oily compounds. Viscous and characteristic in appearance (similar to prepared gelatin), they can be heavy (dense) or have some fluidity (light) (depending on the product).

PART II
INTRODUCTION TO GENERIC DRUG FORMULATION DEVELOPMENT

1. GENERAL STANDARDS OF GOOD MANUFACTURING PRACTICES FOR THE PHARMACEUTICAL INDUSTRY RELATED TO THE DEVELOPMENT OF MEDICINES

Dear Readers, here you will find some general and basic rules about Good Manufacturing Practices, which compliance will help to carry out an adequate development of generic drugs in your industries.

a. About Quality Assurance

I. There must be a Quality Assurance or Guarantee of Quality system that must be coordinated by personnel with the required skills.

II. The company must have a quality policy, which must be duly communicated at all levels of the company.

III. Training or induction (for new personnel) regarding the quality policy will be done according to well-defined written procedures.

IV. There must be a Standard Operating Procedure (SOP) on self-inspections and quality audits.

V. There must be a storage copy for those processes managed electronically.

VI. Access to electronic information systems must be restricted through the use of usernames and passwords that guarantee the confidentiality of the information.

VII. An S.O.Ps control program is required.

VIII. The pilot batches within a formula development process must be prospectively validated prior to their commercialization and at least the first three commercial-scale batches will be validated concurrently.

IX. The documentation generated during product development must be correctly stored.

X. There must be an S.O.P for the release of developed products to the market.

XI. In the event of any deviation in the processes, these will be duly investigated, and the results of this investigation stored.

XII. A Quality Assurance investigation does not apply to preformulation or drug development stages, unless the people responsible for this process detect an anomaly that could impact other products or processes independent of the development carried out.

XIII. Quality Assurance will verify that procedures, instructions, work sheets or any document within the QMS of the development area (or of the development responsible person) is consistent according to the documentary specifications.

XIV. It is recommended to review the documentation of the Q.M.S (Quality Management System) within its period of validity.

XV. Quality Assurance must have a system or means of verification that avoids the use of NON-valid documents.

XVI. Quality Assurance must include within the annual training topics related to Product Development, both for the technical personnel in charge of the process, and for the operational (or administrative) support personnel.

XVII. Quality Assurance training should include topics such as stability studies (accelerated & long term stability and stability for new products under development).

b. About Quality Control

I. There must be an area or sector assigned for the washing and conditioning of materials intended for the preparation of pilot batches (development batches).

II. The Quality Control laboratory must have the necessary equipment, reagents, material and personnel to carry out the analysis of pilot batches in the most optimal way.

III. A correct planning of the representative (or development area) is necessary to be able to indicate to Quality Control, what standards (of active ingredients) must be required to carry out the analysis. With few exceptions, it is only recommended to use Primary Standards for the analysis of batches under development.

IV. All outsourced analysis services must have the respective support documentation (technical agreements or contracts), in case it is required in an audit.

V. Prior to carrying out the pilot batches, the maintenance status (particularly preventive) of the equipment to be involved in the tests must be verified.

VI. All measuring or precision instruments must have their proper calibration.

VII. The measuring (or precision) equipment or instruments must have their calibration status duly labeled.

VIII. All raw material (including packaging material, packaging supplies) used in the development of a pilot lot must have its proper approval.

IX. The raw material (and other) analysis S.O.P must include a section that describes the disposal procedure for those rejected inputs (or raw materials) that have entered as samples for the development of products (that is, they are not part of the normal stock of the pharmaceutical plant).

X. Non-pharmacopoeial analytical methods (or some variation to these) must be validated.

XI. Each dossier or report generated during the pilot batches must have the proper specifications of raw materials, packaging supplies (accessories or inputs) and packaging materials, semi-finished product, finished product; along with their certificates of analysis.

XII. There must be a procedure for the entry, storage and final disposal of those samples from pilot batches generated in the development tests, giving preference to those samples that have maintained their organoleptic, physical and chemical characteristics stable in long-term storage (In what regards to the physical and chemical specifications according to the official pharmacopoeias or to the In-House specifications).

XIII.   Once the product's shelf life of the approved pilot batch (the most stable and that meets commercial requirements) has been determined, it is advisable to store your samples until 1 year after their expiration date.

XIV.   In the case of those samples that will be used as reference substances corresponding to non-pharmacopoeial active ingredients, the appropriate characterization and purity tests of these substances must be carried out.

XV.   The solutions to be used in the analysis must be standardized.

XVI.   The analysis of the pilot batches must have their analytical results. And especially once the final batch is obtained, the records (calculation sheets, worksheets, audit trail, chromatograms) must be kept according to their stabilities.

XVII.   Any error in the documentation (including the documentation of the workbook or preformulation-notebook of the pilot batches) must be corrected with a line, next to the line the correct data, the signature and date of the person that makes the correction (this, according to what is stipulated in the Good Documentation Practices).

XVIII.   Quality control test records shall contain at least the following information:

- Sample identification (name and batch).
- Date.
- Analyst Name.
- Identification of the reference standard.
- Parameters and corresponding conditions.
- Analysis results.Calculations and support documents (see numeral XVI).

2. BASIC PRODUCT DEVELOPMENT CRITERIA

The processes required for product development must be planned, implemented and controlled by the pharmaceutical laboratory, the development department or the person responsible of product development.

The processes for the development of a medicine will be carried out preferably under the following guidelines:

1. The Marketing department (or someone with experience in the pharmaceutical market) will determine the requirements of the product to be developed.

2. Once the product to be developed has been defined, the Marketing department will pass a query or report to a Technical Manager (or his/her designee), in order to verify the feasibility of developing the product in the pharmaceutical laboratory facilities, being the aspects to be considered the following:

a. The product to be developed must be within the scope of the Certificate of Good Manufacturing Practices. That is, I cannot manufacture, for example, a cefuroxime in an area designated for the manufacture of penicillin antibiotics, as cefuroxime is a cephalosporin; this implies that it requires a separate area (even a different plant or facility) for its manufacturing and processing.

b. The product that will be developed (at least in Ecuador) must already have been previously registered in the country with at least three sanitary registries (by competitors in the market or by the same company that wants to develop the product), in what concerns to the same active ingredient and strenght (including here strenghts - concentrations – equal or below to those that I plan to develop). As an example, I want to register an oral solid whose active ingredient is Polyethylene glycol 3350 (osmotic laxative), I first check the medicine database of the health entity (in Ecuador it is ARCSA), and seeing that there are more than 6 manufacturers with the same active, for the same function and pharmaceutical form and in equal or lower concentration, to the one I want to develop; then, I can give the OK (approval) to my development **plan** and continue.

> **This is a documented process, I have not proceeded until here with my pilot scale batches manufacturing.**

In case that you want to register a molecule (active ingredient) in Ecuador for the first time in the country, there is required the complete clinical studies for the approval of the registration (beside all the technical and documented requirements)

c. I must know if my active ingredient (of the product per develop) requires a bioequivalence study or not (this does not apply to injectables)

If somehow that the product per develope requires bioequivalence (it could be in vivo or in vitro (see table according to Ecuadorian regulations), the Management and Commercial deparments must decide whether or not to continue with the development of the product. Generally in Ecuador, bioequivalence tests tend to be very expensive, and the health entity requires that at least the laboratory where the analysis are carried out (in the case of in vitro tests) have the proper accreditation according to the ISO 17025 (until 2022).

LIST OF ACTIVE INGREDIENTS THAT MUST SUBMIT BIOEQUIVALENCE STUDIES IN ECUADOR

NO.	ACTIVE INGREDIENT	STUDY REQUIRED
1	VALPROIC ACID AND ITS SALTS	IN VIVO
2	CARBAMAZEPINE	IN VIVO
3	CYCLOSPORINE	IN VIVO
4	DIGOXIN	IN VIVO
5	ETHAMBUTOL, COMBINATIONS	IN VIVO
6	ETHOSUXIMIDE, COMBINATIONS	IN VIVO
7	EVEROLIMUS	IN VIVO
8	PHENYTOIN SODIUM, COMBINATIONS	IN VIVO
9	GRISEOFULVIN	IN VIVO
10	LEVOTHYROXINE	IN VIVO
11	LITHIUM, SALTS	IN VIVO
12	METHYLDIGOXIN	IN VIVO
13	METHOTREXATE	IN VIVO
14	MYCOPHENOLATE MOFETIL	IN VIVO
15	OXCARBAZEPINE	IN VIVO
16	PROCAINAMIDE	IN VIVO
17	QUINIDINE, COMBINATIONS	IN VIVO
18	THEOPHYLINE, COMBINATIONS	IN VIVO
19	TOLBUTAMIDE	IN VIVO
20	TACROLIMUS	IN VIVO
21	VERAPAMIL	IN VIVO

22	WARFARIN	IN VIVO
23	ALPRAZOLAM	IN VITRO
24	ATENOLOL	IN VITRO
25	BIPERIDEN	IN VITRO
26	CAPECITABINE	IN VITRO
27	CYCLOPHOSPHAMIDE	IN VITRO
28	CLONAZEPAM	IN VITRO
29	CHLORAMBUCIL	IN VITRO
30	GABAPENTIN	IN VITRO
31	LAMIVUDINE	IN VITRO
32	LETROZOL	IN VITRO
33	METFORMIN	IN VITRO
34	PROPRANOLOL	IN VITRO
35	SELEGILINE	IN VITRO
36	TEMOZOLOMIDE	IN VITRO
37	ZIDOVUDINE	IN VITRO
38	LAMIVUDINE + ZIDOVUDINE	IN VITRO

d. After having defined the previous requirements, the person responsible for the formulation begins to see competing products that have a similar formulation (or the same, why not – qualitatively at least -) to the product that I want to develop (- Remember that this book is about generic drugs formulation and product development, but please kindly check your local regulations about medicines, brand or trademark products and laws about patents and copyrights -).

In this case, ie. If I want to develop paracetamol syrup, I can do an internet search for a well-known syrup of a prestigious (identified) brand. For example, I can take a TEMPRA syrup from Laboratorios Bristol as a base (this product is already discontinued in Ecuador, it is just an example).

TEMPRA, in its composition has (-this is only an illustrative example-): Paracetamol (active ingredient). Excipients: purified water, grape flavor, sodium saccharin, FDC red color 40, FDC blue color 1, citric acid, PEG 1450, glycerin, propylene glycol.

e. Once the theoretical formulation of my product (to develop) has been defined, I proceed to consult (speaking in first person as if I were the responsible of the formulation); if in the warehouse of my Pharmaceutical Laboratory I have the substances previously described:

- Paracetamol.

- Grape flavor (take into account that the flavors carry a reference or code, and that this must be mentioned in the formulation that will be submitted in my dossier to the Sanitary Registration Authority or MoH – Ministry of Health-)

- Sodium sacharine.

- Red color FDC # 40.

- FDC Blue Color #1.

- Citric acid.

-PEG 1450.

- Glycerin.

- Propylene glycol

- Water (this must be purified, and the equipment from which it is obtained must have its respective IQ/OQ/PQ). Every batch of water that I use in my tests must be referenced (with a batch number) in my formula. And additionally, the water to be used in my tests must have its proper approval. AVOID at all costs using potable (or drinking) water, no matter how boiled it may be, even in topical preparations. Potable water can be used to wash the materials used in the manufacturing of the pilot batches or for washing the instruments used; BUT, the final rinse will be done with deionized and/or purified water ONLY.

f. In case of not having the raw materials described above, the import request for SAMPLES of the raw materials must be made. To make this request, it is advisable to already have a theoretical formula already defined.

For this we can consult in a HANDBOOK of excipients the recommended percentages. But, to be able to consult in a handbook, or rather the result of the consultation of my handbook will be the determination of the functionality of each one of the excipients.

THIS IS A THEORETICAL FORMULATION

No.	Ingredient	Active Ingredient or Excipient	Functionality	Recommended concentration
1	Paracetamol	API	A.P.I (Active Pharmaceutical Ingredient)	3.200 %
2	Grape Flavor Ref. UVA123	Exc.	Flavoring agent	0.050 % (Concentration would vary according to the flavor once manufactured the pilot batch)
3	Sodium sacharine	Exc.	Sweetening agent	0.025%
4	FDC Red Color #40	Exc.	Colorant	0.012% (Concentration would depend on the formula or because its visual characteristic – organoleptic -)

#				
5	FDC #1 Blue Color	Exc.	Colorant	0.006% (Concentration would depend on the formula or because its visual characteristic)
6	Citric Acid	Exc.	Flavor enhancer	0.300%
7	PEG 1450	Exc.	API solubilizing agent (paracetamol)	14 % (It is soluble in hot or warm water, it may be boiled in direct contact with the product)
8	Glycerin	Exc.	Some sort of flavoring or sweetening enhancer (It has a pleasant slightly sweet taste)	15%
9	Propylene glycol	Exc.	Solvent or carrier (It helps the paracetamol dissolved in the PEG 1450 to mix easily with the glycerin and water)	7%
10	Purified Water	Exc.	Solvent (It would not be a solvent for the active ingredient, since paracetamol only dissolves in PEG 1450)	q.s (quantum satis) (quantity to reach the final volume) (Here in Ecuador in your written formula you must declare the amount of water used, and in this case it would be "in theory" = 60.407 mL of water) In a real situation as the excipients get dissolved or combined between each other; the volume of the bulk before adding (or mixing) all the ingredients from the steps 1 to 9 might not be of 39.593 mL.(%) (39.593 would correspond to the sum of the percentages of the ingredients from the step one (1) to the step nine (9)) Theoretical Water Amount declared in the Dossiers Formula The theoretical amount of water is calculated as follows: 100% - 39,593%= 60.407 mL (%) 60.407 mL is the amount of water that I MUST DECLARE IN THE FORMULA OF THE DOSSIER OF MY FINAL PRODUCT (at least here in Ecuador), but not necessarily this is the quantity of water that will be added until reaching the final volume of 100 mL. For this reason, it is recommendable to work with a liquid-manufacturing tank of known capacity, so, once mixed all the excipients, you can visually realize how many water is needed to reach the final volume of the batch.

g. Once I have my concentrations already established (in theoretical form) as for a 100 mL batch (since we determined a theoretical percentage chart earlier); well, we can make a relationship as for a small 200,000 mL tank (200 liters), or also, we could calculate the necessary amounts for the following volumes to prepare: 100 mL, 1000 mL, and 200,000 mL too.

Let's assume that we do not have ANY of the substances described above and that we need to order or import; more or less the amounts to import would be as follows:

No.	Ingredient	100 mL	1000 mL	200,000 mL	Quantity to Import (check with your Import Representative)
1	Paracetamol.	3.200 %	32.000 g	6400.00 g	Generally, no more than 5 kg are allowed to import as a sample (in Ecuador) (consult this with your local purchasing or import representative), so 5 kg to import would NOT be enough to prepare 200 liters of the batch, therefore, I recommend setting the HIGHEST QUANTITY (or batch volume) to prepare in 100 liters; being the amount to be imported 3,200.00 g of Paracetamol for a 100 litres batch.
2	Grape Flavor Reference. UVA123	0.050 %	0.500 g	100.00 g	50.00 g *
3	Sodium sacharine	0.025%	0.250 g	50.00 g	25.00 g *
4	FDC Red Color #40	0.012%	0.120 g	24.00 g	12,00 g (This amount would be too much in practice, so it is recommended to reduce the colorant, and it should NOT be greater than 4 g per each 200-liter tank.)
5	FDC #1 Blue Color	0.006%	0.060 g	12.00 g	6.00 g *
6	Citric Acid	0.300%	3.000 g	600.00 g	300.00 g *
7	PEG 1450	14 %	140.000 g	28,000.00 g	14,000.00 g (check with your Purchasing Manager if you can buy locally)
8	Glycerin	15%	150.000 g	30,000.00 g	15,000.00 g*
9	Propylene glycol	7%	70.000 g	14,000.00 g	7,000.00 g*
10	Purified water	60,407%	604.070 mL	120,814.00 mL	60,407.00 mL
	Total	100.000%	1000 g	200,000 mL	N/A

Formula per 100 mL and 100 liters (10 batches of 10 liters each can be made)

No.	Ingredient	100 mL	100.000 mL
1	Paracetamol.	3.200 %	3,200.00 g
2	Grape Flavor Ref. UVA123	0.050 %	50.00 g *
3	Sodium Sacharine	0.025%	25.00 g *
4	FDC Red Color #40	0.012%	12.00 g*
5	FDC #1 Blue Color	0.006%	6.00 g *
6	Citric Acid	0.300%	300.00 g *
7	PEG 1450	14 %	14,000.00 g *
8	Glycerin	15%	15,000.00 g*
9	Propylene glycol	7%	7,000.00 g*
10	Purified water	60.407%	60,407.00 mL
	Total	100.000%	N/A

* This asterisk (we saw it on previous pages) indicates that it is necessary to consult or coordinate with the Purchasing Manager, the minimum amount of purchase that can be offered locally, for example, we will generally find (in Ecuador) glycerin and propylene glycol tanks of 200,000 milliliters (200 L), but some sellers can make exceptions and provide us with 50 liter tanks. The same happens with solids, they are usually sold by kilograms (1 kg minimum for purchase); but all this must be carefully discussed with the Purchasing or Import Manager. It is recommended that the Purchasing Manager request samples from the suppliers, all this depends on the company's budget (and the kind of commercial relationship that your company has with the suppliers of raw materials), the important thing here is that this formula to 100 liters (which can be prepared as 5 batches of 20 liters each or 20 batches of 5 liters each), is a THEORETICAL formula. But, it gives us an indication of what raw materials we are going to use.

h. Once the raw material has been received, the pilot batches manufacturing might start.

The process of each batch manufacturing must be recorded in a development log or notebook, which must be coded and numbered as it is part of the Quality Management System (QMS) documentation.

PART III
FORMULAS IN ALPHABETICAL ORDER

A

ACETYLCYSTEINE
POWDER FOR ORAL SOLUTION
200 mg/SACHET

Therapeutic Use: expectorant mucolytic drug.

1. Formula

No.	Ingredientes	Quantity mg/sachet	
1	Acetylcysteine	200.00	mg
2	Aspartame (E951)	26.00	mg
3	Sorbitol powder (E420)	752.84	mg
4	Colorant Sunset yellow (E110)	0.66	mg
5	Orange flavor powder	0.50	mg
6	Colloidal silicon dioxide (E551)	20.00	mg
	Total	1,000.00 (1g)	mg

2. Manufacturing procedure

a. Mix the acetylcysteine with one part (3/4) of the sorbitol and aspartame.

b. Separately, mill (or grind) one part (1/4) of the sorbitol with the colorant and the orange flavor.

c. Mix "a" into "b" and add colloidal silicon dioxide.

If the preparation is not very fluid (powders that tend to stick), try wet granulation, in which instead of using sorbitol you will use sugar (sucrose).

3. Commercial Packaging Presentation Suggested

Box x 30 sachets – each sachet x 1 g. of product

For wet granulation

No.	Ingredients	Quantity mg/sachet
1	Acetylcysteine	66.66 mg
2	Sugar crystals (18-60 mesh)	914.16 mg
3	Sodium sacharine	3.33 mg
4	FD&C Yellow Color No.6	0.66 mg
5	Orange flavor	0.16 mg
6	Aerosil 200 (colloidal silicon dioxide)	20.00 mg
7	Purified water	Lost on drying

1. Manufacturing procedure

a. Charge the acetylcysteine and half the amount of sugar and sodium saccharin into a mixer and mix for 30 minutes.

b. Sieve the pre-mix through a suitable mesh.

c. Load back into the mixer.

d. Add the remaining amount of sugar and colloidal silicon dioxide and mix until the bulk is uniform and homogeneous.

e. Dissolve the colorant in purified water.

f. Continue mixing the powders and slowly add the solution from the previous step (step 'e')

g. When the addition of the solution is complete, manually knead until the bulk is evenly moistened and colored. If necessary, add some amounts of distilled water (in approximately 1 mL increments each).

h. Check that the bulk (mass) is adequate and record (write/register) the total amount of water added. Do not add water in excess.

i. Spread the wet granules on trays and dry at 50 °C until loss on drying (LOD) does not exceed 1%.

j. Let the granules to cool, and then pass them through a granulation machine with a fine mesh.

k. Load the granules from previous step into suitable blender, add flavor and blend until the bulk is homogenous (15 minutes aprox.), passing through a fine mesh if necessary.

l. Place the product in the sachet packer and give each sachet a weight of 3g per sachet.

> Important: Most of the commercial packaging presentations (at least in Ecuador) come within sachets per 1 g. So, a formulation adjustment might be required, base on this calculation:

How do I know that the 66.66 mg of acetylcysteine that my chart states "for a wet granulation process" will give me a concentration in my finished product of 200 mg of acetylcysteine per sachet**?**

0.06666 (66.66 mg of my "wet granulation chart") are in   1000 mg of my bulk

3000 mg (that each sachet has) (How many mg of acetylcysteine would have?)

x= (0.06666 x 3000) / 1000 = 0.19998 g * 1000 = 199.98 mg/sachet or 200 mg/sachet

ACETYLCYSTEINE

600 mg

Effervescent tablets

1. Formula

Theoretical tablet weight: 2000 mg (2g.)

No.	Ingredients	Percentage	
1	Acetylcysteine, crystalline	30.00	%
2	Lactose and polyvinylpyrrolidone (copolymer)	20.00	%
3	Sodium bicarbonate	22.50	%
4	Tartaric acid, powder	17.50	%
5	Polyethylene glycol 6000, powder	3.75	%
6	Aspartame	1.25	%
7	Orange flavor	5.00	%

2. Manufacturing procedure

Mix all the components, pass through a mesh and tablet by direct compression at a maximum atmospheric humidity of 30%.

3. Commercial Packaging Presentation Suggested

Box x 20 tablets + leaflet

ACETYLSALICYLIC ACID

500 mg

Tablets

1. Formula

No.	Ingredients	Quantity mg/tablet	
1	Acetylsalicylic acid, crystalline	500.00	mg
2	Microcrystalline Cellulose PH 101	200.00	mg
3	Polyvinylpyrrolidone K 30	15.00	mg
4	Crospovidone	25.00	mg
5	Magnesium stearate	3.00	mg
Theoretical Weight of the Tablet		743.00	mg

2. Manufacturing procedure

a. Mix

b. Pass the ingredients through the mesh.

c. Compress

Author comment: If the preparation presents drawbacks when tableting, try using the wet granulation method, adding the crospovidone in two parts (one part before granulating and the other after drying), while magnesium stearate is the last excipient before tableting. Use water as a solvent and let the PVP going with the other powders (the bulk); you don't need to add the PVP as a binder solution.

3. Commercial Packaging Presentation Suggested

Box x 40 tablets + insert (leaflet)

ACYCLOVIR

400 mg/5 mL

Oral suspension

Therapeutic Use: oral antiviral drug.

1. Formula

No.	Ingredients	Percentage	
1	Acyclovir	2.00	%
2	Meyprogat (guar gum)	1.00	%
3	Sodium carboxymethylcellulose	0.25	%
4	Liquid sorbitol	20.00	%
5	Methylparaben	0.12	%
6	Propylparaben	0.06	%
7	Citric acid	0.30	%
8	Liquid banana flavor	0.02	%
9	Sodium sacharine	0.08	%
10	Purified water	76.17	%
	Total	**100 mL**	**100%**

2. Manufacturing procedure

a. Place the sodium carboxymethylcellulose in water and allow it to hydrate.

b. Dissolve or rather suspend the acyclovir in the liquid sorbitol (some heating of the sorbitol may be required to facilitate mixing).

c. Separately, dissolve the citric acid and the sodium saccharin in water.

d. Dissolve the methylparaben and propylparaben in a small amount of alcohol (not mentioned in the formula), then add in the alcohol the liquid banana flavor (although water-soluble or sorbitol-soluble banana flavors do exist on the market).

e. Separately dissolve in water the meyprogat (try not to exceed the volume in the theoretical formula – which would be q.s.p 100 mL -).

f. Mix everything (acyclovir with sorbitol, parabens (in alcohol with the flavor), citric acid and saccharin, sodium carboxymethylcellulose already hydrated).

g. Shake vigorously and bring to a volume of 100 mL (remember to use containers whose capacity you can measure or visible know).

h. Take a graduated cylinder and place a sample of the prepared product. Let it be still (do not move it) and the sedimented volume should be small after a few hours.

i. Ensure the stability or homogeneity of your preparation by passing your bulk through the Gaulin homogenizer equipment, at least twice before its primary packaging

3. Commercial Packaging Presentation Suggested

Box x bottle x 125 mL + insert (leaflet) (package the product in an amber bottle)

ALBENDAZOLE

400 mg

Chewable tablets

Therapeutic Use: Antiparasitic (anthelmintic)

1. Formula

No.	Ingredients	mg/tablet	
1	Albendazole	400.00	mg
2	Fast-flo lactose	169.64	mg
3	Cornstarch	40.00	mg
4	Aerosil 200	10.00	mg
5	Sucralose	0.12	mg
6	Tutifruti flavor or a Chewing Gum flavor	0.24	mg
7	Croscarmellose sodium	16.00	mg
8	Di-pac compressible sugar	160.00	mg
9	Magnesium stearate	4.00	mg
10	Purified water	It evaporates during the manufacture, use an adequate quantity until obtaining a compact mass moldable to the hand (yes, to the hand or to the touch)	
Tablet weight		800.00	mg

2. Manufacturing Procedure

a. Mix the albendazole, fast-flo lactose, corn starch, sucralose, and tutifruti flavor or chewing-gum type flavor.

Author comment: Albendazole has a characteristic flavor, so proper trials will be required to obtain a desired flavor profile.

b. Add deionized water to the mixture until an amorphous but moldable mass is obtained.

c. Pass the mixture through a granulator and receive it on racks covered with white wax paper (this is for drying in an oven).

d. Once dry, pass through a granulator and place in a mixer.

e. To this mixture add the Di-Pac (compressible sugar), Aerosil 200, croscarmellose sodium and the magnesium stearate at the end.

f. Tablet with round punches.

3. Commercial Packaging Presentation Suggested

Box x 10 blisters x 10 tablets each + insert (leaflet)

ALPRAZOLAM

0.25 mg

Tablets

Therapeutic Use: Benzodiazepine for the treatment of anxiety disorders.

SCHEDULE IV CONTROLLED SUBSTANCE – IN ECUADOR REQUIRES ALSO
BIOEQUIVALENCE STUDY SUBMISSION TO THE MoH

1. Formula

No.	Ingredients	mg/tablet	
1	Alprazolam	0.25	mg
2	Dicalcium phosphate	80.00	mg
3	Cornstarch	2.00	mg
4	Gelatin	2.00	mg
5	Cornstarch	33.00	mg
6	Propylparaben	0.08	mg
7	Methylparaben	0.08	mg
8	Magnesium stearate	1.00	mg
9	Sodium starch glycolate	1.00	mg
10	Colorant Yellow	0.30	mg
11	Purified Water	q.s.	-
Theoretical Tablet Weight		119.71	mg

2. Manufacturing Procedure

1. Charge items 2 and 5 into a suitable container after sieving through a No. 80 sieve. Mix for 2 minutes.

2. Sieve item 1 through a 60 mesh sieve and add to step 1. (Note: Due to the small amount of item 1, use a suitable dilution method to mix the entire amount- You may take some of the powder from step 1 and mix little by little in a polyethylene bag).

3. Mix for 5 minutes.

4. In a separate container, sieve the bulk (through 80 mesh) and load items 3, 4, 6, 7 and 10; then mix for 2 minutes. Add a sufficient quantity of item 11 to form a suitable, amorphous and mouldable mass. (The color may be previously milled (pulverized) with an adequate part of dicalcium phosphate in order to improve its distribution on the tablet.)

5. Add step 4 into step 3, knead and pass through a granulator until obtaining a homogenous wet granule.

6. Spread the wet granules from step 5 on aluminium trays and dry at 50 °C for 12 hours until the loss on drying test is no more than 2%. Dry for an additional hour, if needed.

7. Pass the dry granules through No. 20 mesh.

8. Sieve items 8 and 9 through a 250 μm sieve screen and add to step 7. Mix for 2 min.

9. Compress using 6 mm. punches. It is recommended that the weight of the tablet should be higher (about 125 mg), compared to the theoretical weight (See formula above).

3. Commercial Packaging Presentation Suggested

Box x 30 tablets + leaflet (insert)

SCHEDULE IV CONTROLLED SUBSTANCE – The label must declare that the product is "BIOEQUIVALENT" (according to Ecuadorian laws)

AMITRIPTYLINE

50 mg

Tablets

Therapeutic Use: tricyclic antidepressant.

1. Formula

No.	Ingredients	mg/tablet	
1	Amitriptyline	50.00	mg
2	Cornstarch	20.00	mg
3	Lactose monohydrate	20.00	mg
4	Dicalcium phosphate	15.00	mg
5	Magnesium stearate	2.00	mg
6	Talc	3.00	mg
7	Cornstarch	20.00	mg
8	Purified water	q.s.	-
Theoretical weight of the tablet		130.00	mg

2. Manufacturing procedure

1. Sieve ingredients 1 to 4 through a 250 μm sieve and load into a suitable blender.

2. In a separate container, load item 2 and add item 8 at 80°C. Mix to form a good paste. Cool to 50 °C.

3. Add step 2 to step 1, and knead until homogenous granules are formed.

4. Spread the wet mass on aluminium trays and dry in an oven with the temperature set on 15°C for 15 hours to a loss on drying (LOD) of not more than 1.5%.

5. Pass dry granules through No. 18 mesh and transfer to a suitable blender.

6. Pass item 5 through a 250 μm sieve and item 7 through a 500 μm sieve; Add to step 5 and mix for 2 minutes.

7. Compress into 130 mg tablets, using a suitable punch.

8. Cover the tablet with a layer of organic base.

3. Commercial Packaging Presentation Suggested

Box x 20 tablets + insert (leaflet)

AMOXICILLIN - CLAVULANIC ACID

250 mg/62,5 mg

Powder for Oral Suspension

Therapeutic Use: Penicillin antibiotic contains beta-lactamase inhibitor.

1. Previous considerations:

a. Take into account the chemical equivalence of the raw materials in order to manufacture the formulation.

EQUIVALENCE:

74.45 mg of Potassium Clavulanate is equivalent to 62.5 mg of clavulanic acid.

287.00 mg of Amoxicillin Trihydrate is equivalent to 250 mg of amoxicillin base (anhydrous).

b. The formula is not expressed as a percentage, it is expressed in grams (g), in this case the formulation has 7 grams which must be packaged in the final container. If a greater amount of powder per bottle is desired, the respective adjustments must be made. The 7 grams when diluted to a volume of 60 mL of suspension will contain the desired concentration of 250 mg of Amoxicillin and 62.5 mg of Clavulanic Acid per 5 mL tablespoon.

c. It is recommended to use the raw material Potassium Clavulanate 1:1 with Syloid, because ( ) is the best raw material to make the formulation.

ATTENTION: As potassium clavulanate is sensitive to oxygen, either from the air or from humidity; it is recommended to nitrogenate the raw material during handling, so that the nitrogen gas displace ambient oxygen and lengthen or improve the stability of the active ingredient.

2. Formula

No.	Ingredients	Each contain	
1	Amoxicillin Trihydrate powder	3.4444	g
	Equivalent to amoxicillin base	3.0000	
2	Potassium clavulanate powder	0.8934	g

		Clavulanic acid	0.7500	
3		Meyprogat (Guar Gum)	0.5000	g
4		Colloidal silicon dioxide	1.5000	g
5		Methylparaben	0.0350	g
6		Propylparaben	0.0175	g
7		Sucralose	0.0747	g
8		Tutifrutti Arom	0.0350	g
9		Calcium Stearate	0.5000	g

MIX THE INGREDIENTS IN THE ORDER STATED IN THE FORMULA
Final volume: 60 mL
Net content: 7.000 g (seven grams)
Glass container, hermetic lid.
Product sensitive to light and humidity.

3. Commercial Packaging Presentation Suggested

Box x bottle (7 g. powder) of 60 mL + insert (leaflet)

AMOXICILLIN - CLAVULANIC ACID

500 mg/125 mg

Tablet

Previous considerations
See the steps: 'a' and 'b' of the previous formula

1. Formula

No.	Ingredients	mg/tablet	
1	Amoxicillin, use compacted amoxicillin trihydrate, with excess	500.00	mg
2	Potassium clavulanate, use potassium clavulanate with avicel (1:1)	125.00	mg
3	Sodium starch glycolate	25.00	mg
4	Colloidal silicon dioxide	30.00	mg
5	Croscarmellose sodium	10.00	mg
6	Talc	10.00	mg
7	Magnesium stearate	5.00	mg
Theoretical weight of the tablet		705.00	mg

2. Manufacturing Procedure

1. Dry item 1 at 45 °C for 2 hours.
2. Dry items 3, 5, 6 and 7 at 80 °C for 4 hours.
3. Sieve items 1 to 7 through #40 mesh, load contents into a rectangular blender for tableting processes, and mix for 30 minutes.
4. Compress the mixture of the step 3 using 16 mm punches and a hardness of 6 to 7 kPa.

5. Break the tablets from the previous step by passing them through 2.5 mm mesh sieves through a granulator.

6. Transfer the granules from the previous step to a mixer and add items 6 and 7 for 15 minutes.

7. Compress with 19×9 mm punches.

8. Coat the tablets with HPMC organic coating

3. Commercial Packaging Presentation Suggested

Box x 20 tablets + insert (leaflet).

ATORVASTATIN CALCIUM

10 mg

Tablet

Therapeutic Use: Antilipidemic agent.

1. Formula:

No.	Ingredients	mg/tablet	
1	Atorvastatin, use atorvastatin calcium trihydrate	11.000	mg
2	Calcium carbonate	36.000	mg
3	Lactose monohydrate	65.000	mg
4	Microcrystalline cellulose PH 102	30.000	mg
5	Polyvinylpyrrolidone K 30	3.000	mg
6	Polysorbate 80	0.400	mg
7	Croscarmellose sodium	4.000	mg
8	Magnesium stearate	0.600	mg
9	Purified water	q.s.	-
Theoretical Weight of the Tablet		150.000	mg

2. Manufacturing procedure

1. Sieve atorvastatin calcium trihydrate, calcium carbonate, lactose monohydrate, and microcrystalline cellulose PH 102 through a 0.500 mm stainless steel sieve.

2. Dissolve polyvinylpyrrolidone and polysorbate 80 in purified water (50 °C) with slow stirring until clear. Cool the solution to 30 °C. This is the granulation solution.

3. Knead the powdered mixture with granular solution to obtain the desired granules.

4. Dry the granules to a loss on drying of 2%.

5. Pass the dry granules through the #16 mesh.

6. Sieve the sodium croscarmellose and magnesium stearate through a 0.500 mm mesh.

7. Load the sieved granules from step 5 and the powder mix from step 6 into a suitable blender. Mix for 1 minute.

8. Compress into 150 mg tablets, using 12 mm punches.

Author comment: For a 20mg concentration, compress 300mg into 15mm punches.

9. Prepare a solution of hypromellose and polyethylene glycol 4000 in a mixture of purified water and 95% ethanol. Hold overnight for complete gelation.

10. Add the talc and titanium dioxide and mix to obtain a uniform coating dispersion.

11. Proceed to coat the tablet.

3. Commercial Packaging Presentation Suggested

Box x 30 tablets + insert (leaflet)

AZITHROMYCIN ORAL SUSPENSION

250 mg/5mL

Therapeutic Use: Broad spectrum antibiotic.

1. Formula:

No.	Ingredients	100 mL	
1	Azithromycin dihydrate	5.00	g.
2	Citric acid	2.00	g.
3	Sodium citrate	5.00	g.
4	Crospovidone	9.00	g.
5	Liquid orange flavor	0.05	g
6	Polyoxyl 40 hydrogenated castor oil CAS Number: 61788-85-0 DL 50 oral rat: 20g/Kg	0.50	g
7	Purified Water (q.s.)	78.45	mL

2. Manufacturing procedure

a. Mix the azithromycin, citric acid, sodium citrate, and crospovidone in half of the purified water.

b. In the PEG hydrogenated castor oil (I recommend Eumulgin – with the characteristics or specifications mentioned in the formula-) dissolve the liquid orange flavor and add 20 mL of purified water with constant stirring.

c. Add the mixture of step 'b' in 'a' and reach to final volume with water.

Author comment:

1. If the preparation tends to separate into phases (or does not appear homogeneous), it is recommended to add meyprogat little by little and in proportion to the amount of crospovidonein the formula. That is, if I add meyprogat (or guar gum) I remove what is added (from meyprogat) to the crospovidone to maintain the percentage concentration. You could also try Avicel RC 591 if you don't have meyprogat (guar gum).

2. Given the characteristic of the active ingredient with respect to taste, the use of a Magnasweet type taste masker (monoammonium glycyrrhinate alone or combined) is recommended.

3. Commercial Packaging Presentation Suggested

Box x 30 mL bottle + insert

BROMHEXINE ELIXIR

8 mg/5mL

Therapeutic Use: Mucolytic, expectorant.

1. Formula:

Ingredients	100 mL	400 mL
Bromhexine hydrochloride	0.1600 g	0.6400 g
Alcohol	9.7800 mL	39.1200 mL
Sorbitol	30.0000 mL	120.0000
Propylene glycol	10.0000 mL	40.0000 mL
Glycerin	20.0000 mL	80.0000 mL
Hydrochloric acid *	0.00083 mL	0.00332 mL
Sodium sacharine	0.0100 g	0.0400 mL
Peach flavor	0.04800 mL	0.1920 mL
Purified water	30.00117 mL	120.00468 mL

* Hydrochloric acid (HCl) will be previously prepared as a solution of equivalent concentration (or Normality) (0,1N). The prepared solution will have a concentration of 0.00083 mL of HCl/mL. Then this solution will be placed in a suitable container, where the pilot batch will be manufactured and the final volume will be of 100 mL.

2. Manufacturing procedure

1. The alcohol is placed in a tank.
2. Bromhexine hydrochloride is subsequently placed.

3. Propylene glycol is added.

4. Then sorbitol is added.

5. Glycerin is added.

6. 0.1N hydrochloric acid is added.

7. Sodium saccharin is added.

8. Finally, the peach flavor is added (previously dissolved in alcohol or propylene glycol – it will depend on the raw material).

9. Complete the volume of the bulk with purified water.

3. Commercial Packaging Presentation Suggested

Box x 240 mL bottle + insert (leaflet).

C

CEFUROXIME AXETIL TABLETS

500 mg

Therapeutic Use: Second generation oral cephalosporin antibiotic.

General Considerations about This Product

a. Before starting with the development or planning of this product, check the scope of its Certificate of Good Manufacturing Practices. This is because antibiotics such as cefuroxime require a special area for their preparation; it must not be manufactured in a plant where penicillin drugs are produced.

b. The present formulation corresponds to the cores (uncoated tablets) of a cefuroxime axetil formulation.

c. It is recommended to use the raw material Cefuroxime Axetil DC (Direct Compression) over other raw materials, such as Cefuroxime Axetil Amorphous Compacted. Based on previous tests, Cefuroxime Axetil Direct Compression has better properties and gives better results.

d. This formula shows a theoretical case where the raw material: Cefuroxime axetil Direct Compression (DC) has an equivalence where 1.1500 g of Cefuroxime Axetil DC contains 0.500 mg of Cefuroxime base.

e. This formula is made up of the active ingredient, a 2% disintegrant, a 0.5% dissolution-enhancing agent, and a 0.15% lubricant. **There is no diluent (or filler for the tablet)** because the active ingredient itself has good compactibility.

f. In case you want to obtain a more precise equivalence compared to the one shown in this assay, consult your raw material supplier in order to do the respective calculation with the CoA (certificate of analysis) of the raw material (Cefuroxime Axetil DC-Direct Compression).

1. Formula:

No.	Ingredientes	Grams por Tablet
1	Cefuroxime Axetil Direct Compression (DC)	1.1500 g.
2	Croscarmellose sodium	0.0240 g.
3	Sodium lauryl sulfate	0.0060 g.
4	Magnesium stearate	0.0018 g.
Theoretical weight of the tablet		1.1818 g.

2. Manufacturing procedure

Mix the ingredients in the order described in the formula. Compress the tablets with capsular punch. The resulting product will be an off-white tablet. Protect it from light and moisture. It would be not so probable that a double compression will be required; if this happens increase the croscarmellose sodium and sodium lauryl sulfate (in this case the tablet weight would no longer be 1.1818 g) in order to let them to work extragranularly.

CIPROFLOXACIN TABLETS
1000 mg

Therapeutic Use: Antibacterial agent from the group of fluoroquinolones.

<u>Considerations regarding this product</u>

a. This formulation corresponds to the core (uncoated tablets) of a formulation of Ciprofloxacin tablets 1000 mg.

b. Based on prolonged release products, it is recommended to make a cover using: Hydroxypropylmethylcellulose (HPMC), polyethylene glycol, succinic acid (as a pH stabilizer), and titanium dioxide.

c. Take into account the addition of water at the time of making the granules so that there is not an excess of it with a consequent decrease in the concentration of the active ingredient.

1. Formula:

No.	Ingredients	Grams per tablet
1	Ciprofloxacin HCl	1.164 g.
2	Microcrystalline Cellulose PH 101	0.030 g.
3	P.V.P (polyvinylpyrrolidone)	0.014 g.
4	Colloidal silicon dioxide	0.005 g.
5	Magnesia stearate	0.034 g.
6	Croscarmellose sodium	0.020 g.
7	Purified Water	q.s * + * Got lost on drying + Quantity necessary to obtain an amorphous and moldable mass
Theoretical tablet weight		1.267 g.

2. Manufacturing procedure

The procedure is by wet granulation.

1. Mix the Ciprofloxacin, Microcrystalline Cellulose PH 101, PVP, and half of the Croscarmellose Sodium. Add water until you get an amorphous and moldable mass.

2. Add the second part of the croscarmellose, colloidal silicon dioxide, and magnesium stearate.

3. Compress, double compression is most likely required. Use a capsular punch that is wider than longer. This is because the tablet tends to have a height that could make it difficult to blister.

4. Don't forget to coat your tablet.

3. Commercial Packaging Presentation Suggested

Box x 10 tablets + insert

CLOTRIMAZOLE TOPICAL SOLUTION

1%

Therapeutic Use: Topical antifungal.

1. Formula:

No.	Ingredients	Each 100 mL
1	Clotrimazole (previously micropulverized)	1.00 g.
2	Polyoxyl 40 hydrogenated castor oil (preferable: Eumulgin)	29.93 g.
3	Methylparaben	0.05 g.
4	Propylparaben	0.02 g.
5	Ethanol 96°	34.00 g.
6	Purified water	35.00 g.
Final Volume		100.00 mL

2. Manufacturing procedure

a. Place the Eumulgin in a little bit of alcohol (1/4 approx.) and add little by little (with constant stirring) all the clotrimazole (previously micropulverized).

b. Separately, dissolve the parabens in hot water.

c. Incorporate the dissolved parabens into the mixture from step A.

d. Add to the mixture from step c the rest of the alcohol

e. Complete with warm water. Be careful as the preparation has high alcohol content.

3. Commercial Packaging Presentation Suggested

Bottle x 30 mL + insert

CODEINE SYRUP

10 mg/5mL.

Therapeutic Use: Antitussive.

1. Formula:

Ingredients	100 mL	1000 mL
Codeine phosphate	0.2000 g	2.0000 g
Alcohol	8.0000 mL	80.0000 mL
Rum aroma	0.0074 mL	0.0740 mL
Methylparaben	0.1000 g	1.0000 g
Propylparaben	0.0500 g	0.5000 g
FDC Red Color #40	0.0090 g	0.0900 g
Sugar crystals	58.8000 g	588.0000 g
Vanilla	0.2500 mL	2.5000 mL
Purified water	32.5836 mL	325.8360 mL

2. Manufacturing procedure

a. In ¾ of the water dissolve the sugar and prepare the syrup. Add the codeine and shake until dissolved.

b. Once the sugar is dissolved, dissolve apart the parabens in a small amount of water (must be hot). Add once dissolved to the previous mixture.

c. Separately, in alcohol, place the rum aroma, the vanilla (essence) and add to the previous mixture.

d. Add the colorant.

e. Complete with purified water.

3. Commercial Packaging Presentation Suggested

Box x 240 mL bottle + insert

CODEINE TABLETS

30 mg

Therapeutic Use: Antitussive.

1. Formula:

No.	Ingredients	mg/tab
1	Codeine phosphate hemihydrate	30.000 mg
2	Microcrystalline Cellulose PH 200	120.900 mg
3	Croscarmellose sodium	2.700 mg
4	Colloidal silicon dioxide	1.350 mg
5	Magnesium stearate	0.130 mg
Theoretical tablet weight		155.080 mg

2. Manufacturing procedure

Mix the ingredients in the stated order, taking into account the amount of added lubricant (magnesium stearate) as it can influence the dissolution tests.

3. Commercial Packaging Presentation Suggested

Box x 10 tablets + insert

COLCHICINE TABLETS

0,6 mg

Therapeutic Use: Antigout agent

1. Formula

Ingredients	1 tablet
Colchicine	0.000625 g
Magnesium stearate	0.000200 g
Avicel PH 200 LM	0.085000 g
Fast-flo lactose	0.013000 g
P.V.P (polyvinylpyrrolidone)	0.001250 g
Ac-di-sol (Croscarmellose sodium)	0.001250 g
Purified Water	0.004000 mL

2. Manufacturing Procedure

1. Colchicine, fast-flo lactose, Avicel PH 200 LM and PVP are weighed and then mixed.

2. Add the water and mix.

3. Add the acdisol.

4. Magnesium stearate is incorporated. Mix for no more than 3 minutes.

Compress with flat round punches.

3. Commercial Packaging Presentation Suggested

Box x bottle x 60 tablets + insert

D

DIMENHYDRINATE TABLETS

100 mg

Therapeutic Use: prevention of nausea, vomiting and dizziness.

1. Formula

No.	Ingredients	1 tablet
1	Dimenhydrinate	100.000 mg
2	Microcrystalline cellulose PH 101	127.870 mg
3	Starch	20.080 mg
4	Croscarmellose sodium (AC-DI-SOL preferable)	1.040 mg
5	P.V.P (polyvinylpyrrolidone)	0.520 mg

6	Alcohol (Enough quantity to obtain an amorphous mass that is hand-moldable) (Be careful as dimenhydrinate is soluble in alcohol)	q.s.
7	Magnesium stearate	0.490 mg
	Tablet weight	250.00 mg

2. Manufacturing procedure

a. Mix the dimenhydrinate, half of the microcrystalline cellulose PH 101, all of the starch, half of the AC-DI-SOL, the P.V.P.

b. Add the alcohol.

c. Knead until you obtain an amorphous mass that is moldable to the touch.

d. Pass the mixture through a granulator and receive it on trays covered with wax-paper.

Place in an oven and dry, BUT BE CAREFUL WITH THIS, since the mixture can "melt" or become liquid if the drying time and temperature are not well determined.

d. Once the mixture is dry, proceed to break the granulate the mixture (pass it through the granulator).

e. Add the remainder of the microcrystalline cellulose PH 101, and the rest of the croscarmellose sodium, and mix.

f. Add the magnesium stearate.

g. Compress (no coating is required).

3. Commercial Packaging Presentation Suggested

Box x 20 tablets + insert

E

EMULSION (LINIMENT) FOR SPRAINS, BUMPS, AND BRUISES (METHYL SALICYLATE, MENTHOL, CAMPHOR)

3:1:1

Pharmaceutical dosage form: Liniment.

1. Formula:

Ingredients	100 mL
Methyl salicylate	15.0000 g
Mentol	5.0000 g
Camphor	5.0000 g
Solid Vaseline	6.5900 g
Tween 80	1.9800 g
Arlacel 80	1.3100 g
Cetyl alcohol	1.6400 g
Synchrowax BB4 (synthetic beeswax)	0.6500 g
Carbopol 940	0.1400 g
Triethanolamine	0.1300 g
Purified water	6.5600 mL

IMPORTANT

During the preparation of this product, it is recommended to use a full-face mask (with a visor and pre-filters), since the substances in the product could cause damage to both nasal and oral mucosa. This situation does NOT extend for any reason to the product already packaged; this safety measure is only recommended during the product preparation.

2. Manufacturing procedure

a. As the methyl salicylate is a liquid product, proceed to dissolve the menthol and camphor in it.

b. Heat Vaseline, tween 80, arlacel 80, cetyl alcohol, and Synchrowax BB4 (alternative to natural beeswax) in a suitable container.

c. Dissolve the mixture from step A in the mixture from step B (all fats and oils already melted).

d. In water dissolve carbopol and let it hydrate.

e. Once the carbopol is hydrated, add the triethanolamine and the mixture will gelify slightly giving the appearance of a cloudy and somewhat viscous solution.

f. Add the mixture from step D to the mixture from step E. At this point your product will turn a bright white color.

g. Shake constantly until incorporated.

h. Let the product still, there should be no phase separation.

i. Proceed to pack.

3. Commercial Packaging Presentation Suggested

White PET bottle (without box) of 240 mL + Insert (it could be digital or attached to the bottle by some suitable means).

ERYTHROMYCIN TABLETS

500 mg

Class of medication: Macrolide antibiotic.

1. Formula:

Each core (uncoated tablet) contains:

Ingredients	g/tablet
Erythromycin Ethylsuccinate	0.5850 g
Avicel pH 200	0.0664 g
Magnesium stearate	0.0030 g
Talc	0.0270 g
Sodium lauryl sulfate	0.0420 g
Acdisol	0.0100 g

2. Manufacturing procedure

a.   Mix the erythromycin ethylsuccinate, avicel pH 200, and acdisol.

b.   Then add sodium lauryl sulfate and mix.

c.   Add the talc and mix.

d.   Add magnesium stearate and mix for a maximum of 3 minutes.

e. Compress (may require double compression, if so, start with small and thin capsular tablets; and later, once broken, give them the desired weight).

f. Coat the tablet.

Author's comment: Please pay attention to the dissolution rate of the tablet.

3. Commercial Packaging Presentation Suggested

Box x 10 strips of 10 tablets each + insert.

L

LANSOPRAZOLE DISPERSIBLE TABLETS

20 mg

Class of medication: Proton-pump inhibitor.

Dispersible tablets: Those that are previously dissolved in water before ingestion.

1. Formula:

No.	Ingredients	mg/tablet
1	Lansoprazole	20.00 mg
2	Calcium Lactate	175.00 mg
3	Calcium glycerophosphate	175.00 mg
4	Sodium bicarbonate	250.00 mg
5	Aspartame	0.50 mg
6	Colloidal silicon dioxide	12.00 mg
7	Cornstarch	15.00 mg
8	Croscarmellose sodium	12.00 mg
9	Dextrose anhydrous	10.00 mg
10	Peppermint flavor	3.00 mg
11	Maltodextrin	3.00 mg
12	Manitol	3.00 mg
13	Pregelatinized starch	3.00 mg

2. Manufacturing procedure

a. Pass all the ingredients through a 250 μm mesh and mix in a suitable blender.
b. Compress to a weight of **682 mg**, using round punches, flat on both sides and 15 mm in size.

3. Commercial Packaging Presentation Suggested
Box x 10 blisters of 10 tablets each + insert

MOUTHWASH (ALCOHOL-FREE)

(Type: Zero)

Pharmaceutical dosage form: Solution

1. Formula:

Ingredients	Quantity
Chlorhexidine gluconate	0.1200 mL
Methyl salicylate	0.00900 g
Eumulgin HRE 40	0.41061 g
Menthol	0.09000 g
Eucalyptol	0.22477 g
Thymol	0.00450 g
Propylene glycol	3.53882 mL
Sorbitol	4.77876 mL
Novamint Mint flavor	0.03539 mL
Sodium sacharine	0.05309 g
FDC Green Color #3	0.00018 g
Purified Water	q.s.

VERY IMPORTANT CONSIDERATIONS

The concentration of chlorhexidine of this formula is ideal to eliminate 99.99% of bacteria in mouth cavity, and I have had the opportunity to test this on different occasions.

In the event that browning of the gums or teeth is observed due to the use of chlorhexidine gluconate, it is recommended to add some type of fluorinated compound typical of toothpastes to prevent this situation.

If fluoride or fluorinated products are difficult to import into your country, trials can be performed using 50% benzalkonium chloride (instead of chlorhexidine gluconate). Where 0.025900 g of 50% benzalkonium chloride equals 0.012500 g in 100 mL of mouthwash.

However, benzalkonium chloride can be irritating to the tonsils and pharynx (the use of this substance in gargles would not be recommended for any reason) and its use in mouthwashes is subject of debate (the use of benzalkonium chloride in mouthwashes is not very widespread).

2. Manufacturing Procedure

a. In eucalyptol, dissolve methyl salicylate, timol, mentol and mint liquid flavor.

b. Add the Eumulgin Hre 40 (which we have previously used in this book)

c. Add propylene glycol to the previous mixture.

d. Add sorbitol.

e. Add some water and the resulting mixture will be transparent and with scarce or almost none opalescence.

f. Apart in water dissolve chlorhexidine gluconate and add to the previous mixture.

g. Also in water dissolve sodium saccharin, and the dye, and add to the previous mixture.

h. Complete volume with purified water.

3. Commercial Packaging Presentation Suggested

240 mL transparent PET bottle (without box) with metal pilfer-proof cap (the insert can be digital or attached to the bottle by some suitable mean).

O

OMEPRAZOLE DISPERSIBLE TABLETS

20 mg

Class of medication: Proton-pump inhibitor.

Dispersible tablets: Those that are previously dissolved in water before ingestion.

Instructions for use: Place a tablet in a glass with approximately 30 mL of water and shake gently until dissolved. After this, ingest the preparation.

Once the tablet is dissolved (if you don't like so much drinking water) you can add a lemonade type juice or orange juice in case you want to taste another flavor. PLEASE DO NOT DISSOLVE THE TABLET IN MILK.

1. Formula:

No.	Ingredients	mg/tablet
1	Omeprazole	20.00 mg
2	Calcium lactate	175.00 mg
3	Calcium glycerophosphate	175.00 mg
4	Sodium bicarbonate	250.00 mg
5	Aspartame	0.50 mg
6	Colloidal silicon dioxide	12.00 mg

7	Cornstarch	15.00 mg
8	Croscarmellose sodium	12.00 mg
9	Dextrose anhydrous	10.00 mg
10	Peppermint Flavor	3.00 mg
11	Maltodextrin	3.00 mg
12	Mannitol	3.00 mg
13	Pregelatinized starch	3.00 mg

2. Manufacturing procedure

a. Pass all the ingredients through a 250 µm mesh and mix in a suitable blender.
b. Compress to a weight of 682 mg, using round punches, flat on both sides and 15 mm in size.

3. Commercial Packaging Presentation Suggested

Box x 28 tablets + leaflet

P

PARACETAMOL CHEWABLE TABLETS

300 mg

1. Formula:

I. Paracetamol (pulverized)........... 300 g
    Sugar pulverized   .... 600 g
    Crospovidone .... 550 g
    Strawberry flavor (powder).......... 60 g
II. Polyvinylpyrrolidone K 30........... 60 g
    Ethanol al 96%......................... 425 g

2. Manufacturing procedure

a) Granulate the mix I with solution II, pass through a screen and compress with medium compression force. Add magnesium stearate before tableting.

Tablet weight: 1,620.00 mg; diameter: 20 mm; shape: flat round punch; hardness (in Newtons): 111 N.

3. Commercial Packaging Presentation Suggested

a. Box x 1 blister of 10 tablets + leaflet.
b. Box of 10 blisters of 10 tablets each + leaflet.
c. Box x 50 blisters of 10 tablets each + leaflet (
)

PARACETAMOL TABLETS

1000 mg

1. Formula:

g. / tablet:
1.1000 g. Paracetamol compacted (Direct Compression or DC) equivalent to 1.000 g de paracetamol
0.0220 g. Croscarmellose sodium
0.0165 g. Aerosil 200
0.0055 g. Magnesium stearate
Mix ingredients in the order described and compress by direct compression.

GENERAL CONSIDERATIONS

- For the development of this product it is necessary to take into account that the tablet will have a considerable height. It is necessary to compress with a scored capsular punch that allows obtaining tablets that are wider than taller.

- Prior to the formulation study, consider whether you have the appropriate formats to blister the product.

- Aerosil 200 absorbs moisture and prevents increase of tablet hardness during storage.

PASSIFLORA-VALERIAN

(Uncoated tablet)

Use: Natural mild sedative.

1. Formula:

Ingredients	General Name	1 tablet	200 tablets
Passionflower dry extract	-	0.04000 g	8.00000 g
Valerian root dry extract	-	0.05000 g	10.00000 g
Avicel PH 101	Microcrystalline cellulose	0.07900 g	15.80000 g
Alcohol	Alcohol (INN)	0.02127 g	4.25400 g
Lactose	Lactose (D.C.I)	0.05000 g	10.00000 g
Magnesium stearate	Magnesium stearate (D.C.I)	0.00090 g	0.18000 g
Acdisol	Croscarmellose sodium	0.00900 g	1.80000 g
Parteck S.I 400	Sorbitol DC (Directly Compressible)	0.01000 g	2.00000 g

Raw material used in the formulation
- **Passionflower Extract**

Botanical name: Passiflora Incarnata

Extraction Type: water/ethyl alcohol (water/grain alcohol)

Plant part: flower (dried)
Active ingredient: flavonoids – expressed by vitexin concentration.

For **further information about vitexin in what regards to other medicinal and pharmacological properties**, please kindly check the following article:

Miao He, Jia-Wei Min, Wei-Lin Kong, Xiao-Hua He, Jun-Xu Li, Bi-Wen Peng,
A review on the pharmacological effects of vitexin and isovitexin,
Fitoterapia,
Volume 115,
2016,
Pages 74-85,
ISSN 0367-326X,
https://doi.org/10.1016/j.fitote.2016.09.011.
(https://www.sciencedirect.com/science/article/pii/S0367326X16304488)
Abstract: Vitexin and isovitexin are active components of many traditional Chinese medicines, and were found in various medicinal plants. Vitexin (apigenin-8-C-glucoside) has recently received increased attention due to its wide range of pharmacological effects, including but not limited to anti-oxidant, anti-cancer, anti-inflammatory, anti-hyperalgesic, and neuroprotective effects. Isovitexin (apigenin-6-C-glucoside), an isomer of vitexin, generally purified together with vitexin, also exhibits diverse biological activities. Latest research has suggested that vitexin and isovitexin could be potential substitute medicines for diversity diseases, and may be adjuvants for stubborn diseases or health products. This review summarized recent findings on various pharmacological activities and associative signalling pathways of vitexin and isovitexin to provide a reference for future research and clinical applications.
Keywords: Vitexin; Isovitexin; Pharmacological activities; Traditional Chinese medicine

- **Valerian Root Extract**

Botanical name: Valeriana officinalis

Extraction type: water ethyl alcohol (water/grain alcohol)

Plant part: root.

Active ingredient: valerenic acid.

2. Manufacturing procedure

a. Granulate the ingredients of the preparation (through wet granulation process), dividing only the PARTECK -S.I 400 into two equal parts (one is wet granulated and the other part is placed after the granulation process).

b. Magnesium stearate is extragranular.

You can coat the tablet with a simple layer (film type, the simplest, without the need for any pigment).

At the time that I developed this formulation, there were commercial presentations of competing products that came in the form of dragees, but the formula that I propose here (as I remarked previously) plenty fullfil its pharmacological function.

UNDECYLENIC ACID SOLUTION

25%

Therapeutic Use: topical antifungal.

Note: The solution is useless in nails; it is indicated for skin only.

1. Formula

Ingredients	For a batch of 100 mL	For a batch of 1,600 mL (1.6 L)
Undecylenic Acid [*]	22.755g=25mL	364.08g =400 mL
Isopropyl Palmitate	75 mL	1200 mL

[*] Undecylenic Acid:
at 25°C it has a density of about: 0.9102/cc

2. Manufacturing Procedure

Gradually mix the undecylenic acid into the isopropyl palmitate with constant agitation.

Comment:

Take the necessary precautions when preparing the product, since the mixture is oily and any spillage in your work area, if not handled well, could cause falls.

3. Commercial Packaging Presentation Suggested

Box x 120 mL bottle + insert (leaflet) + applicator brush

APPENDIX

ADDITIONAL PHARMACEUTICAL MANUFACTURING CONSIDERATIONS

MANUFACTURING PRACTICE CONSIDERATIONS IN LIQUID MANUFACTURING

The manufacture and control of oral solutions and oral suspensions presents some unusual problems not common to other dosage forms. Although bioequivalency concerns are minimal (except for products in which dissolution is a rate-limiting or absorption-determining step, as in phenytoin suspension), other issues have frequently led to recalls of liquid products. These include microbiological, potency, and stability problems. In addition, because the population using these oral dosage forms includes newborns, pediatrics, and geriatrics, who may not be able to take oral solid dosage forms and who may have compromised drug metabolic or other clearance function, defective dosage forms can pose a greater risk if the absorption profiles are significantly altered from the profiles used in the development of drug safety profiles (Niazi,S. 2004).

Facilities.

The designs of the facilities are largely dependent on the type of products manufactured and the potential for cross-contamination and microbiological contamination. For example, the facilities used for the manufacture of over the counter (OTC) oral products might not require the isolation that a steroid or sulfa product would require. However, the concern for contamination remains, and it is important to isolate processes that generate dust (such as those processes occurring before the addition of solvents). The HVAC (heating, ventilation, and air-conditioning) system should be validated just as required for processing of potent drugs. Should a manufacturer rely mainly on recirculation rather than filtration or fresh air intake, efficiency of air filtration must be validated by surface and air sampling. It is advisable not to take any shortcuts in the design of HVAC systems, as it is often very difficult to properly validate a system that is prone to breakdown; in such instances a fully validated protocol would need stress testing — something that may be more expensive than establishing proper HVAC systems in the first place. However, it is also unnecessary to overdo it in designing the facilities, as once the drug is present in a solution form; cross contamination to other products becomes a lesser problem. It is, nevertheless, important to protect the drug from other powder sources (such as by maintaining appropriate pressure differentials in various cubicles) (Niazi,S. 2004).

Equipment

Equipment should be of sanitary design. This includes sanitary pumps, valves, flow meters, and other equipment that can be easily sanitized. Ball valves, the packing in pumps, and pockets in flow meters have been identified as sources of contamination. Contamination is an extremely important consideration, particularly for those sourcing manufacturing equipment from less developed countries; manufacturers of equipment often offer two grades of equipment: sanitary equipment, and equipment not qualified as sanitary and offered at substantial savings. All manufacturers intending to ship any product subject to U.S. Food and Drug Administration (FDA) inspection must insist on certification that the equipment is of sanitary design.

To facilitate cleaning and sanitization, manufacturing and filling lines should be identified and detailed in drawings and standard operating procedures. Long delivery lines between manufacturing areas and filling areas can be a source of contamination. Special attention should be paid to developing standard operating procedures that clearly establish validated limits for this purpose. Equipment used for batching and mixing of oral solutions and suspensions is relatively basic. These products are generally formulated on a weight basis, with the batching tank on load cells so that a final volume can be made by weight; if you have not done so already, consider converting your systems to weight basis. Volumetric means, such as using a dipstick or a line on a tank, are not generally as accurate and should be avoided where possible. When volumetric means are chosen, make sure they are properly validated at different temperature conditions and other factors that might render this practice faulty. In most cases, manufacturers assay samples of the bulk solution or suspension before filling. A much greater variability is found with those batches that have been manufactured volumetrically rather than those that have been manufactured by weight. Again, the rule of thumb is to avoid any additional validation if possible (Niazi,S. 2004).

The design of the batching tank with regard to the location of the bottom discharge valve often presents problems. Ideally, the bottom discharge valve is flush with the bottom of the tank. In some cases, valves — including undesirable ball valves — are several inches to a foot below the bottom of the tank. This is not acceptable. It is possible that in this situation the drug or preservative may not completely dissolve and may get trapped in the "dead leg" below the tank, with initial samples turning out subpotent. For the manufacture of suspensions, valves should be flush. Transfer lines are generally hard piped and are easily cleaned and sanitized. In situations where manufacturers use flexible hoses to transfer product, it is not unusual to see these hoses lying on the floor, thus significantly increasing the potential for contamination. Such contamination can occur through operators picking up or handling hoses, and possibly even through operators placing them in transfer or batching tanks after the hoses had been lying on the floor. It is a good practice to store hoses in a way that allows them to drain, rather than coiling them, which may allow moisture to collect and be a potential source of microbial contamination.

Another common problem occurs when manifold or common connections are used, especially in water supply, premix, or raw material supply tanks. Such common connections can be a major source of contamination (Niazi,S. 2004).

Raw Materials

The physical characteristics, particularly the particle size of the drug substance, are very important for suspensions. As with topical products in which the drug is suspended, particles are usually very fine to micronized (to <25 microns). For syrup, elixir, or solution dosage forms in which there is nothing suspended, particle size and physical characteristics of raw materials are not that important. However, they can affect the rate of dissolution of such raw materials in the manufacturing process. Raw materials of a finer particle size may dissolve faster than those of a larger particle size when the product is compounded. Examples of a few oral suspensions in which a specific and well-defined particle-size specification for the drug substance is important include phenytoin suspension, carbamazepine suspension, trimethoprim and sulfamethoxazole suspension, and hydrocortisone suspension. It is therefore a good idea to indicate particle size in the raw material specification, even though it is meant for dissolving in the processing, to better validate the manufacturing process while avoiding scale-up problems (Niazi,S. 2004).

Compounding

In addition to a determination of the final volume (on weight or volume basis) as previously discussed, there are microbiological concerns, and these are well covered in other chapters in this book. For oral suspensions there is the additional concern of uniformity, particularly because of the potential for segregation during manufacture and storage of the bulk suspension, during transfer to the filling line, and during filling. It is necessary to establish procedures and time limits for such operations to address the potential for segregation or settling as well as other unexpected effects that may be caused by extended holding or stirring. For oral solutions and suspensions, the amount and control of temperature is important from a microbiological as well as a potency aspect. For those products in which temperature is identified as a critical part of the operation, the batch records must demonstrate compliance using control charts. There are some processes in manufacturing in which heat is used during compounding to control the microbiological levels in the product. For such products, the addition of purified water to make up to final volume, the batch, and the temperatures during processing should be properly documented. In addition to drug substances, some additives æ such as the most commonly used preservatives, parabens are difficult to dissolve, and require heat (often to 80°C). The control and verification of their dissolution during the compounding stage should be established in the method validation. From a potency aspect, the storage of product at high temperatures may increase the level of degradants. Storage limitations (time and temperature) should be justified. There are also some oral liquids that are sensitive to oxygen and that have been known to undergo degradation. This is particularly true of the phenothiazine class of drugs, such as perphenazine and chlorpromazine. The manufacture of such products might require the removal of oxygen, as by nitrogen purging. In addition, such products might also require storage in sealed tanks, rather than in those with loose lids. Manufacturing directions provided in this book are particularly detailed about the purging steps, and these should be closely observed (Niazi,S. 2004).

Microbiological Quality

Microbiological contamination can present significant health hazards in some oral liquids. For example, some oral liquids, such as nystatin suspension, are used in infants and immunocompromised patients, and microbiological contamination with organisms (such as Gram-negative organisms) is not acceptable. There are other oral liquid preparations such as antacids in which Pseudomonas sp. contamination is also objectionable. For other oral liquids such as cough preparations, contamination with Pseudomonas sp. might not present the same health hazard. However, the presence of a specific Pseudomonas sp. may also indicate other plant or raw material contamination and often points to defects in the water systems and environmental breaches; extensive investigations are often required to trace the source of contamination. Obviously, the contamination of any preparation with Gram-negative organisms is not desirable (Niazi,S. 2004).

In addition to the specific contaminant being objectionable, such contamination would be indicative of a deficient process as well as an inadequate preservative system. For example, the presence of a Pseudomonas putida contaminant could also indicate that *P. aeruginosa*, a similar source organism, is also present.

Because FDA laboratories typically use more sensitive test methods than industry, samples of oral liquids in which manufacturers report microbiological counts well within limits may be found unacceptable by the federal laboratories. This result requires upgrading the sensitivity of testing procedures (Niazi,S. 2004).

Oral suspensions

Liquid products in which the drug is suspended (not in solution) present some unique manufacturing and control problems. Depending on the viscosity, many suspensions require continuous or periodic agitation during the filling process. If delivery lines are used between the bulk storage tank and the filling equipment, some segregation may occur, particularly if the product is not viscous. Procedures must therefore be established for filling and diagrams established for line setup prior to the filling equipment. Good manufacturing practice would warrant testing bottles from the beginning, middle, and end of a batch to ensure that segregation has not occurred. Such samples should not be combined for the purpose of analysis. In-process testing for suspensions might also include an assay of a sample from the bulk tank. More important at this stage, however, may be testing for viscosity (Niazi,S. 2004).

Product Specifications

Important specifications for the manufacture of all solutions include assay and microbial limits. Additional important specifications for suspensions include particle size of the suspended drug, viscosity, pH, and in some cases, dissolution. Viscosity can be important, from a processing aspect, to minimize segregation. In addition, viscosity has also been shown to be associated with bioequivalency.

pH may also have some meaning regarding effectiveness of preservative systems and may even have an effect on the amount of drug in solution. With regard to dissolution, there are at least three products that have dissolution specifications. These products include phenytoin suspension, carbamazepine suspension, and sulfamethoxazole and trimethoprim suspension. Particle size is also important, and at this point it would seem that any suspension should have some type of particle size specification. As with other dosage forms, the underlying data to support specifications should be established(Niazi,S. 2004).

Process Validation

As with other products, the amount of data needed to support the manufacturing process will vary from product to product. Development (data) should have identified critical phases of the operation, including the predetermined specifications that should be monitored during process validation. For example, for solutions, the key aspects that should be addressed during validation include ensuring that the drug substance and preservatives are dissolved. Parameters such as heat and time should be measured. In-process assay of the bulk solution during or after compounding according to predetermined limits is also an important aspect of process validation. For solutions that are sensitive to oxygen or light, dissolved oxygen levels would also be an important test. Again, the development data and the protocol should provide limits. As discussed, the manufacture of suspensions presents additional problems, particularly in the area of uniformity. The development data should address the key compounding and filling steps that ensure uniformity. The protocol should provide for the key in-process and finished product tests, along with their specifications. For oral solutions, bioequivalency studies may not always be needed. However, oral suspensions, with the possible exception of some of the over-the-counter antacids, usually require a bioequivalency or clinical study to demonstrate their effectiveness. Comparison of product batches with the biobatch is an important part of the validation process. Make sure there are properly written protocol and process validation reports and, if appropriate, data for comparing full-scale batches with biobatch available during FDA inspection (Niazi,S. 2004).

Stability

One area that has presented a number of problems is ensuring the stability of oral liquid products throughout their expiry period. The presence of water or other solvents enhances all reaction rates: Because fluids can contain a certain amount of oxygen, the oxidation reactions are also enhanced, as in the case of vitamins and the phenothiazine class of drugs. Good practice for these classes of drug products should include quantitation of both the active and primary degradant. There should be well-established specifications for the primary degradant, including methods of quantitation of both the active drug and degradant. Because interactions of products with closure systems are possible, liquids and suspensions undergoing stability studies should be stored on their side or inverted to determine whether contact of the drug product with the closure system affects product integrity. Other problems associated with inadequate closure systems are moisture losses that can cause the remaining contents to become superpotent and microbiological contamination (Niazi,S. 2004).

Packaging

Problems in the packaging of oral liquids have included potency (fill) of unit dose products and accurate calibration of measuring devices such as droppers, which are often provided. For unit dose solution products the label claim quantity within the limits described should be delivered. Another problem in the packaging of oral liquids is lack of cleanliness of the containers before filling. Fibers and even insects often appear as debris in containers, particularly in the plastic containers used for many of these products. Many manufacturers receive containers shrink-wrapped in plastic to minimize contamination from fiberboard cartons, and many manufacturers use compressed air to clean the containers. Vapors, such as oil vapors, from the compressed air have occasionally been found to present problems, and it is a good practice to use compressed gas from oil-free compressors (Niazi,S. 2004).

Dear Reader, if you have reached this point in the book, I thank you so much.

ABOUT THE AUTHOR

Francisco De La Torre Quiñónez, Ecuadorian Chemist and Pharmacist, is a professional with considerable experience in Formula Development, Implementation and Design of Pharmaceutical Validation Strategies, and in Sanitary Registration of Medicines in General.

After his Ebook entitled: <u>Oral Pharmaceutical Dosage Solids Formulation Manual (Spanish Edition)</u>, released in 2022. Francisco De La Torre brings us in 2023, this work entitled: **"Generic Drugs Formulation Manual: Basic Principles of New Product Development"**; in which he covers not only the development of formulations in oral solid dosage forms, but also brings us formulations of semi-solid, liquid and semi-liquid dosage forms, in what regards to general medicines.

As a "Plus +" to this work, the author brings us a formula of a natural product developed by him years ago, which has been subjected, tested, and approved to a pharmacological study on animals.

The book: **"Generic Drugs Formulation Manual: Basic Principles of New Product Development"**, is an interesting text for all professionals related to the pharmaceutical industry, and also constitutes the cornerstone or starting point for the implementation of a unit or a development deparment in those companies that wish to have this type of process within their industries.

If you want to consult with the author of this book about the formulas of this treaty, or, if you are in

Ecuador and want to publish your book, do not hesitate and do it with us.

Write us to the following email:

publicationsandcompany@gmail.com

-    EDLT PUBLICATIONS -